D1061385

Surviving Antarctica

David N. Thomas

Foreword by Ray Mears

FIREFLY BOOKS

Thanks Gerhard

Previous page:
The distinctive tracks of emperor penguins as they march across the Antarctic landscape.
This page:
Strong and freezing winds sweep over the Southern Ocean.

A FIREFLY BOOK

Published by Firefly Books Ltd. 2007

First Printing

Publisher Cataloging-in-Publication Data (U.S.)
Thomas, David N., 1962–
Surviving Antarctica / David N. Thomas.
[96] p. : col. photos ; cm.
Includes bibliographical references and index.
ISBN-13: 978-1-55407-294-1
ISBN-10: 1-55407-294-8
1. Antarctica – Description and travel.
I. Title. 919.8/9 dc22 G863.T566 2007

Library and Archives Canada Cataloguing in Publication
Thomas, David N. (David Neville), 1962–
Surviving Antarctica / David N. Thomas.
Includes bibliographical references and index.
ISBN-13: 978-1-55407-294-1
ISBN-10: 1-55407-294-8
1. Natural history — Antarctica. 2. Zoology — Antarctica. 3. Antarctica. I. Title.
QH84.2.T48 2007 508.98'9 C2007-900368-0

Edited by Vicky Paterson
Text designed by David Mackintosh

Published in the United States by
Firefly Books (U.S.) Inc.
P.O. Box 1338, Ellicott Station
Buffalo, New York 14205

Published in Canada by
Firefly Books Ltd.
66 Leek Crescent
Richmond Hill, Ontario L4B 1H1

Printed in Dubai

contents . . .

Foreword by Ray Mears

When asked, people will tell you willingly of their fear of being bitten by snakes in the jungle, or of the horror of dying of thirst in the desert, yet rarely mention fear of the cold. But when faced by it, cold generates a terror unique in the human psyche. I have often wondered whether this is some vestigial memory of the ice ages. Whatever it is, we are warm-blooded, warmth-loving creatures. The very thought of the remote southern continent, a seeming wasteland of ice and snow sculpted by unrelenting winds, is one of desolation. A hell that would fit the Norse conception, cold and without cheer, a place of despair and abandon.

Great indeed were the pioneers who first set south to explore this wasteland, and mighty were their reports on the scale, majesty and challenges of this continent. The explorers of Antarctica have provided us with southern sagas every bit the equal of their northern cousins. Tales of endurance, heroism, endeavor punctuated by bitter failures and frozen corpses. These tales captured the imagination of the world. Even today, would-be heroes head south to add the sheen of Antarctic gloss to their reputations, but of course the pioneers who truly set out into the unknown were the real and worthy heroes.

The greatest tale of Antarctica has to be the journeys to the southern pole by Scott and Amundsen. Each in their way would unwittingly provide 50% of the solution for coping with life in Antarctica. Amundsen applied proven technology from the peoples of the Arctic to make his journey. But his most valuable contribution was the wisdom of choosing professionals, specialists in given fields, to man his team, tempered by a thoroughly no-nonsense mental approach to the task. Scott, in a strange irony, gave his life pioneering the approach to southern travel, which has largely come to be the way of the South. Ahead of his age, he employed mechanized transport and modern fabric clothing and arrived with the intention of exploring the scientific secrets of the South.

Since those early days, the Antarctic has not lost any of its mystery, ferocity or wonder, but it has been more fully mapped and understood. Through humankind's persisting curiosity, new techniques have been developed to cope with Antarctica's unremitting nature. Of course, new technologies have played their part as well, both in coping with the environment and its exploration. The personal challenge of venturing this far south has not changed, and in blizzards or during the Antarctic winter, the sense of remoteness and personal isolation are just as they always were. But today, people travel there to gain a better understanding of our planet and particularly the way in which we as a species are impacting our biosphere.

Despite the apparent toughness of this southern continent, it is in reality an extremely fragile, highly specialized ecosystem. In this fascinating book David Thomas takes us to Antarctica in a very personal way, detailing the nature of Antarctic living and the kind of research carried out by scientific teams and, importantly, highlighting the environmental pressures facing Antarctica. Dispelling the fear of the Antarctic, the reader is won over to the value of protecting this realm from environmental harm and the custody of future generations.

BLOG:

[I] AM FULLY ANTARCTIC WINTER SURVIVAL FIELD TRAINED. I CAN SAW BLOCKS OF SNOW, BUILD ICE HOLES

... The trick is to keep your gloves and hat on all the time, sleep with your drinking water bottle so it doesn't freeze, but not get it mixed up with your pee bottle! I'm proficient at melting snow for tea, have climbed into and out of a crevasse, and have my car, quad bike and snowmobile licenses.

• • •

Are you
ready to go?

Earth's beautiful south

"THE END OF THE WORLD," "DESOLATE FROZEN WASTELANDS," "the most inhospitable place on Earth" and "a forsaken place" are just some of the common descriptions of Antarctica. Why would you want to go there? Well, talk with anyone who has made the journey and, as well as tales about freezing temperatures and frostbite, they will wax lyrical about its beauty. Snow glistening as if millions of diamonds had just been scattered on the surface. Intense blue ice carved into wonderful sculptures. The midnight sun. Icebergs as big as cathedrals. The awesome silence and crystal clear waters. And all this before they talk about the wildlife they encountered.

Icebergs as large as cathedrals never cease to impress the Antarctic visitor.

Of course, it's not just the beauty of the Antarctic that has drawn explorers, adventurers and, more recently, tourists to venture south. It's a fantastic place to study, and the easiest way of getting there is as a scientist. The vast majority of visitors to the continent and the surrounding ocean are researchers, there to explore fundamental questions about our planet. Experts in climatology, atmospheric science, geophysics and oceanography work alongside biologists, geologists and chemists. Even astrobiologists, those studying extraterrestrial life, use the Antarctic to test their theories. Some fly in for just a few days or weeks. The ones who really get the Antarctic bug, and are fortunate to secure funding, build up a career that enables them to come back again and again.

SCIENCE IN ANTARCTICA Just as science is a complex web of knowledge, so scientists rarely work alone. If you're a scientist working in Antarctica, you are one part of a much larger support system. Each scientist is associated with a base or research ship, each financed and run by national organizations. These in turn belong to the Scientific Committee on Antarctic Research (SCAR), a group of 28 nations with active research programs in Antarctica, as well as representatives from another 10 to 15 countries. The governments of these nations recognize the importance of the Antarctic for Earth as a whole, and take an active role in SCAR activities. Collectively, SCAR is responsible for facilitating and coordinating high-quality international scientific research in Antarctica. And the work that comes out of it helps determine how Antarctica fits into the Earth's system.

WHAT DOES IT COST? Despite these layers of multinational organization, you still have to apply for funds to put your scientific ideas into action. This doesn't happen overnight, and proposals and the full cost of your science project have to be submitted to national funding organizations several years before anything happens. Be it a proposal to study plankton in the ice-covered Antarctic waters or the effects of global climate change, your research proposal needs to include details of cost. And it isn't cheap. First there are the costs to get the team down there. Then there are the funds needed to do the actual work, including collecting samples, analysing them and writing up the results. Each researcher and technician will need a salary, both for the fieldwork and any follow-up work back home. Every last little bit of your scientific equipment and polar clothing has to be listed, as well as any survival training you need and the necessary pre-expedition medical checkups. Not as easy as it sounds three or four years before you intend to start the project.

When we arrive on board all the laboratories are completely empty, so when it has been less bumpy we have found our 60 or so boxes of chemicals and equipment ... and begun to set up our laboratory. These were delivered to Bremerhaven ... transported to Cape Town to be on board when we arrived. It is always a big relief to see that the stuff has made it. 

If there's one thing that will help your request for funds get approved, it's if you show how the planned work fits into the larger scheme of research in Antarctica. For this reason, scientists from different countries often combine activities to make stronger cases, and this is where SCAR comes in. Again this takes time, so before a research proposal is even written, there may have already been a decade or more of preliminary work invested in it. With only 33 research stations and a handful of research ships, competition is fierce. When the positive answer arrives, there is a real sense of elation and the real planning, rather than just the wistful dreaming, can start.

Preparation

Packing takes on a whole new meaning when preparing for Antarctica. Once you're there, you can't just run out to the store for a forgotten set of batteries, or phone up a supplier for a replacement bit of equipment. Absolutely everything you could possibly need has to be taken with you. A golden rule for scientists is to always take spare parts. If you think you'll need 10 collecting bottles, take 20. If you do run out of things, be prepared to improvise. One expedition ran out of glass sample bottles, so they drank as much soda pop as they could and used the screw-capped bottles instead so that their precious samples could be preserved and stored.

But pack too much, and you'll pay the Earth in shipping costs. It's a difficult trade-off between what you absolutely need and what would simply be desirable. Deciding months before you go makes it even more difficult. For ship-based expeditions, chemicals and equipment may have to be stowed on board six months or more in advance. Imagine having to pack for your summer holiday three months before you go. It can sometimes be a big surprise to open up a box you packed ages ago and wonder why on earth you packed a particular item in the first place.

Everything needed for an Antarctic expedition has to be stowed away in containers many months before.

IT'S THE LITTLE THINGS THAT COUNT

Don't get completely distracted by the essential gear you'll need. In among your thermals and scientific equipment, bring lots of reading material, a music collection and a stack of DVDs. They make all the difference when the nearest movie theater or bar is weeks away. Make room for your favorite cookies or chocolate, too, as well as mementos from home.

As for the bulkier stuff, skis and climbing gear are a must for some, whereas for others it's the latest camera and video equipment (with plenty of spare batteries). And it's just as high a priority to pack the high-factor sunblock and sunglasses to combat the skin-frying ultraviolet radiation. Good-quality polar clothing is essential and is often supplied by the national Antarctic organization funding the work. You really get a sense that the trip is coming close when trying on the bulky overalls, thick, padded jackets and well-insulated boots. Fur-lined gloves and hats keep out the wind and freezing temperatures, and it's worth taking time to make sure things fit and are comfortable – after all, these are what will keep you alive out there.

Conducting fieldwork can be physically demanding.

An array of clothing layers is needed to keep out the cold.

TEST EVERYTHING – EVEN YOU Although there are doctors and other medical personnel at most of the bases on Antarctica and on all the research ships, it's important everyone arrives medically fit. The level of pre-expedition medical tests will depend on how long you go for and how remote your work is. Even for short stints you'll need a series of blood, heart- and lung-function tests, as well as intensive dental checkups. Tests become more rigorous for longer stays. The last thing you want is for your trip (and the work of your colleagues) to be compromised by ill health, and it's important to make sure you're up to the physically demanding strains of working at cripplingly low temperatures.

For the longer trips, those up to several years, some countries will scrutinize your emotional aptitude to be away for long periods at a time. Sharing a relatively small living space with the same people day after day, far from home and under extreme conditions, won't suit everyone. Any problems need to be identified before you go, or they could jeopardize a whole expedition.

Field laboratory – even here sophisticated chemistry can be carried out.

BLOG:

... There is much about ship life to get used to, such as how does a 5-foot (1.5 m) tall person (me) get in and out of the top bunk? Answer – carefully! ... Also, meals at 7:30 am, 11:30 am and 5:30 pm take a bit of getting used to, but I am a student and we eat whenever food is presented.

TRAINING MATTERS Depending on what work you'll be doing, there will be a host of different seminars and survival training courses to attend before you go. These will cover everything from first aid and what to do when caught in a blizzard to ways of minimizing your impact on the pristine Antarctic landscape. You'll learn about Antarctica's wildlife and how to repair a broken-down snowmobile. And you'll hear about the everyday highs and lows of going south from people who have already been. Teams spending a long time on the continent will practice working at low temperatures and in snow and ice. There will be survival training on how to build shelters in the snow and how to traverse crevasse-ridden landscapes. These are key times for team building, where you start to work together and see if everyone is up to the challenge.

IT ALL HAS TO COME BACK The fragile environment of Antarctica needs to be protected from all the scientific and leisure interest. The Antarctic Treaty was set up to do just that, designating the area and its associated ecosystems as a natural reserve, devoted to peace and science. Specifically when planning a research trip, it rules that:

> *"The protection of the Antarctic environment and dependent and associated ecosystems and the intrinsic value of Antarctica, including its wilderness and aesthetic values and its value as an area for the conduct of scientific research, in particular research essential to understanding the global environment, shall be fundamental considerations in the planning and conduct of all activities in the Antarctic Treaty area."*

In all the packing and preparation, you have to keep in mind that what you take to the Antarctic you have to bring back. This is a relatively recent development, but is now strictly adhered to and is in fact law. Wastes are strictly controlled, separated into organics, burnable, glass, metals, batteries, hazardous waste and waste oil. These have to be shipped out and disposed of properly outside the Treaty area. Even human waste has to be collected. Some field groups working in very remote places may be given permission to discharge their urine and water for washing (known as gray water) under strict guidelines, but they need to ask. Everything else has to be brought back to base. Since these are mostly located on coastal sites, the sewage and gray water from the base is treated in water treatment plants and discharged into the sea under strictly controlled protocols, which are becoming increasingly more rigorous with time.

THE LAW The Antarctic Treaty, a concise document running only two pages, was signed in 1959 and came into force in 1961. It covers the whole area south of 60°S latitude. Its main objectives are simple and quite unique in international relations:

- *To demilitarize Antarctica, to establish it as a zone free of nuclear tests and the disposal of radioactive waste, and to ensure it is used for peaceful purposes only.*
- *To promote international scientific cooperation in Antarctica.*
- *To set aside disputes over territorial sovereignty.*

All the waste produced from all the bases on the Antarctic needs to be collected and removed.

An aircraft at the South Pole, where the flags represent some of the 12 original signatories to the Antarctic Treaty.

There are now 44 signatories to the Treaty (27 with voting rights) that meet each year. Over the past 45 years, more than 200 recommendations and five separate international agreements have been negotiated. Together with the original Treaty, these provide the set of rules governing all activities in Antarctica.

Ready to go

Eventually the packing is finished and the documentation complete. Everything is now ready to be shipped off by plane or by sea. A major part of the preparation is finished. All that's left is to get yourself to Antarctica. This might be a two- or three-week trip by ship, or more commonly a flight with a considerably shorter journey time of less than a day. However, as you start your final leg of the journey, the anticipation is immense. All the planning, training, paperwork, packing and mental preparation have been geared to this point. And a network of logistical teams has been working to get you there. It's about now you start to wonder exactly how you'll react to one of the most remote and hostile parts of the planet. Although the working conditions are hardly the same as for the early explorers, the thought is quite sobering.

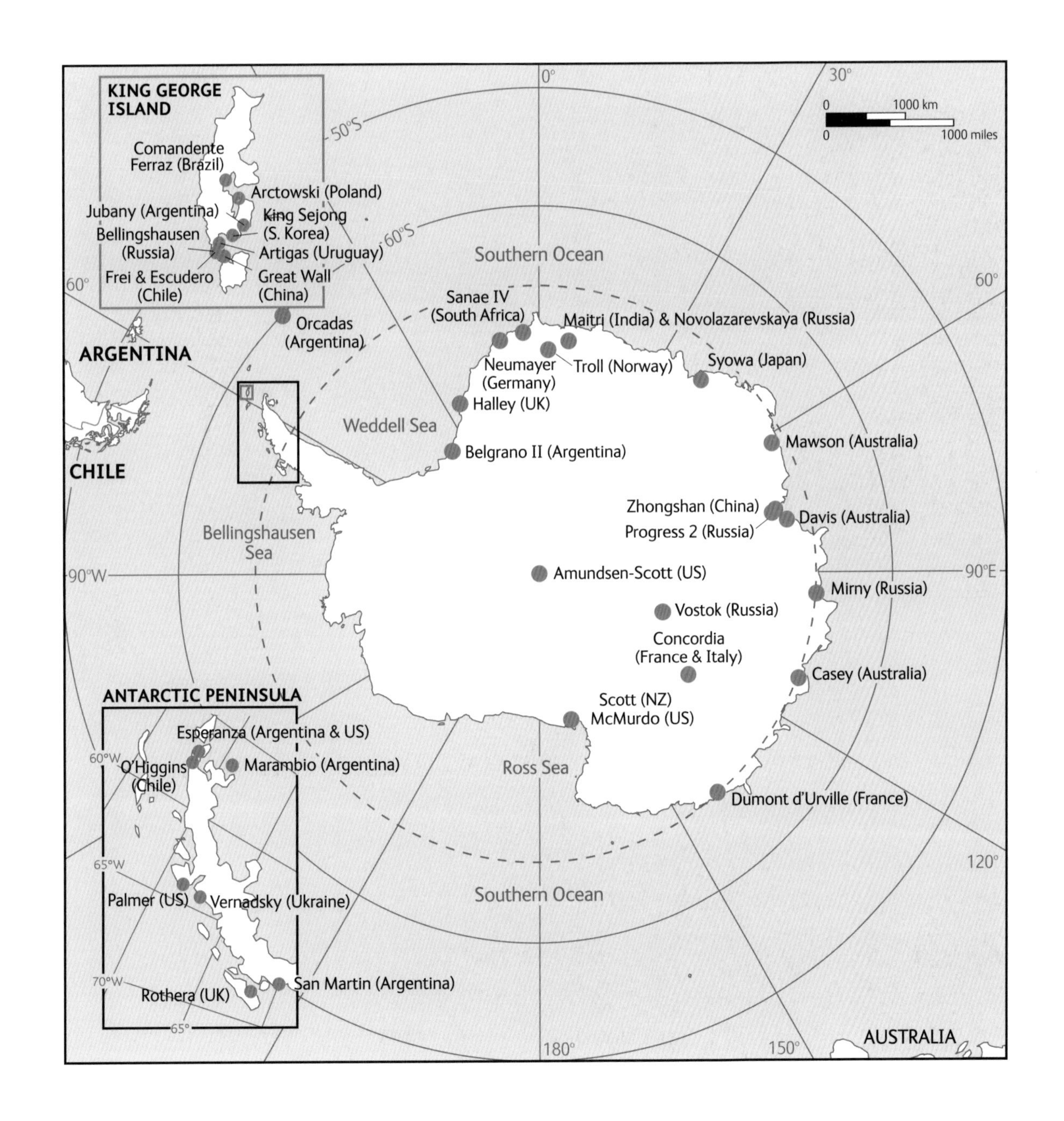

The main year-round Antarctic stations and the countries that operate them.

BLOG:

IT'S THE WEIRDEST SNOW AS WELL, LIKE FAKE SNOW.

It's so cold it doesn't stick together at all so there is no resistance when you walk through it, it takes no effort, it just blows up in the air ... And it's like shoveling polystyrene chips in a gale, there's absolutely no weight to it ...

You've landed

What is Antarctica?

ANTARCTICA IS SIMPLY ANOTHER NAME for the Antarctic continent, which is about 5.4 million square miles (14 million sq km) of rock mostly covered in ice. That's one-and-a-half times the size of the United States. People sometimes refer to the Antarctic, which is the whole area south of the latitude 60°S and also includes most of the Southern Ocean. The Antarctic Circle, at 66.5°S, is the farthest latitude from the south pole where there is at least one day that the sun does not fall below the horizon in summer or does not rise above it in winter. Being in the southern hemisphere, winter lasts from May to September while December to February are the summer months.

The Southern Ocean is bordered by the surface, eastward-flowing Antarctic circumpolar current (ACC), which transports the greatest amount of water of any of the ocean currents on Earth. About 25 million years ago, the Drake Passage opened up between the tip of the Antarctic Peninsula and the tip of South America, enabling the ACC to flow around the continent. The effect of the ACC on the heat balance and weather patterns around Antarctica is one of the main reasons why the Antarctic is so cold. When the ACC formed, it effectively cut off the Southern Ocean and Antarctica from the rest of the world, and this barrier helps explain why many species of animals, plants and single-celled organisms that evolved during the intervening time are only found south of the ACC.

The Antarctic continent is an ice-and-glacier-covered mountainous landscape.

Antarctica in relation to other landmasses.

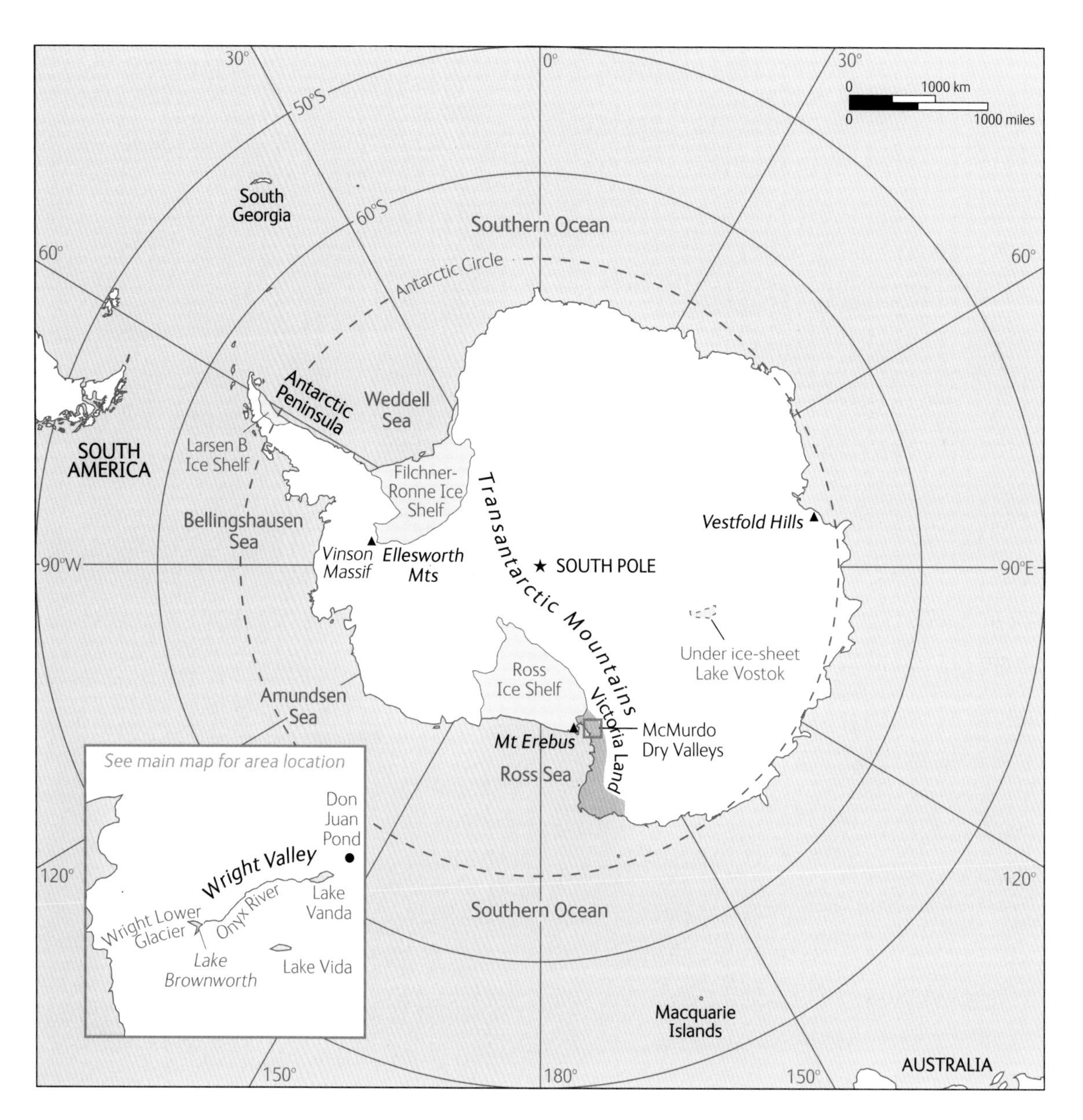

The Antarctic continent.

The vast expanse of the Antarctic is peppered with huge mountains and sweeps of icy plains. It is crudely split into three regions: East Antarctic, West Antarctic from which the Antarctic Peninsula, the third part, extends, stretching up toward South America. East and west are split by the Transantarctic Mountain Range, a 1988-mile (3200 km) long divide, roughly the distance between Australia's north and south coasts. Its highest point, Vinson Massif in the western Ellsworth Mountains, reaches four times the height of the Empire State Building at 16,863 feet (5140 m).

90% OF THE ICE ON THE PLANET A large proportion of Earth's water is locked away in solid form, such as pack ice, lake ice, river ice, snow cover, glaciers, ice caps and ice sheets, as well as in the frozen ground (including permafrost). Collectively this solid water is known as the cryosphere. It is fundamental to the energy budgets, climate and ocean circulation of the planet. There is no doubt that the greatest proportion of the cryosphere is found in the Antarctic. Studies that go on here to understand how this frozen world is maintained are paramount to our understanding of the earth system as a whole.

There is so much ice covering Antarctica that in some places only the mountaintops are visible.

This rocky continent is draped in a massive ice cap — a collection of thick sheets of ice laid down each year by falling snow. Snow settles on the surface, but cannot melt in the constantly cold temperatures. As the snow gets buried under fresh snow, its crystals change. They eventually form ice crystals that fuse into a new sheet of solid ice. As the ice gets buried further, it changes its properties, and at great depths of thousands of feet the pressure caused by the weight of the ice sheets on top can actually cause the ice to melt back into a liquid. About 75% of the world's freshwater is locked up in this ice cap, and it accounts for 90% of all the ice on the planet. You can see why climatologists are concerned about it possibly melting. If it all melted, global sea levels would rise by 215 feet (65 m).

The ice is on average 7545 feet (2300 m) thick, reaching a maximum of 15,650 feet (4770 m) over East Antarctica. Whole mountains are hidden underneath, although peaks known as nunataks do poke out in places. The ice actually pushes the underlying rocks down so that if the ice were removed the rocks would rise slowly by about 3280 feet (1000 m). There is no ice sheet covering the Antarctic Peninsula, although much of that mountainous terrain is still covered by massive glaciers. The West Antarctic Ice Sheet is thinner than that covering East Antarctica, and it is mostly settled on rock that is actually below sea level. This contrasts with the East Antarctic Ice Sheet, which forms a dome over rock above sea level.

KEEPING IT COLD Solar radiation will warm the snow and ice, but that depends on how much of the incoming energy is actually absorbed. Albedo is the term given to the ratio of radiation reflected and the amount hitting a surface. White surfaces have high albedos, reflecting energy very effectively, whereas dark bodies are poor reflectors and absorb energy. Soils and water bodies have low albedos compared to white snow. Warm, wet snow has a lower albedo than dry, fresh snow. High albedo is critical to maintaining the ice sheets. The East Antarctic Ice Sheet receives the maximum monthly input of solar energy of anywhere on Earth, but since most of this is reflected back, it remains the coldest place on the planet.

SLOW-GROWING ANCIENT ICE Although we imagine Antarctica to be a snow-swept place, the atmosphere is so dry that not much snow falls at all, only about 1–4 inches (2–10 cm) each year. It's just that the snow does not melt. It is, in fact, so dry it can be called a desert. Consequently, the ice sheets grow very slowly indeed. The colossal ice forms we see today have taken hundreds of thousands of years to create. Since the ice builds in annual layers, it captures an archive of changing climate conditions. Each year, the chemistry of the snow and air trapped between the crystals are laid down, and these records can tell us how conditions have changed over the years.

CLUES FROM ICE CORES Scientists, such as those involved in the European Project for Ice Coring in Antarctica (EPICA) at Dome C on the East Antarctic Ice Sheet, take cores through this ancient ice to date the ice and perform fine-scale chemical analyses of the gases and chemicals trapped inside. One core extracted by the EPICA group is about 10,728 feet (3270 m)

White snowy surfaces reflect solar energy keeping them cold, whereas dark surfaces, like the sea, absorb energy and warm up.

long, representing a climate record for the past 900,000 years. This sample and other cores from the Antarctic, as well as from the Greenland Ice Sheet in the Arctic, are the best evidence we have that today's concentration of greenhouse gases such as carbon dioxide and methane are the highest they have been for more than 650,000 years.

UNDER THE ICE Lake Vostok was the first lake to be discovered under the Antarctic ice sheets during the drilling of the 11,975-foot (3650 m) long Vostok ice core in 1996. The core stopped just 328 feet (100 m) above the surface of the lake, and an international group of scientists are trying to devise a way to take samples of the lake water without contaminating it. At about 5405 square miles (14,000 sq km) and up to 1640 feet (500 m) deep, the lake is as big as Lake Ontario. Since then, another 145 smaller lakes have been found scattered across the continent, each a treasure trove to scientists. The water they contain was covered by ice millions of years ago, and may contain microbial life isolated from the rest of the planet all that time. These isolated forms could help us understand how life first evolved.

ICE ON THE MOVE Although heavy, ice sheets are really super-glaciers and are by no means static. They flow outward from their highest points. In places the ice sheets are actually moving on a kind of slurry caused by the meltwater formed at the base of the ice sheet, and in other places the base of the ice sheet remains frozen but still moves slowly over the underlying terrain. The flow is halted when it meets a solid barrier such as the Transantarctic Mountains,

which effectively act as a dam. The ice sheets back up against the barriers and breach gaps between the mountains as glaciers. The majority of the ice sheets, however, eventually flow into the sea where they form what are known as floating ice shelves. The largest of these is the Ross Ice Shelf, which at about 188,032 square miles (487,000 sq km) is the size of France, and it has an average thickness of 984 feet (300 m).

Eventually the ice shelves break up, disintegrating further to form icebergs. One of the most dramatic ice-shelf breakups in recent years was the collapse of Larsen B on the eastern side of the Antarctic Peninsula in 2002. Within just over a month, 1255 square miles (3250 sq km) of the shelf – three times the size of Hong Kong – broke off from the continent and disintegrated into thousands of icebergs that drifted off into the Weddell Sea. The amount of ice released in this short time was in the order of 806 billion tons (720 billion metric tons).

The glaciers are moving, albeit slowly, sometimes cascading as icefalls.

Eventually the glaciers and ice shelves break up into weird and wonderful icebergs.

Large glaciers can also flow into the sea traveling as fast as 1.2 miles (2 km) per year. The front edge of the glacier, or tongue, floats on the surface of the sea, then breaks off to produce icebergs up to several hundred square miles large – the size of a large city such as London or New York – although many are much smaller, the size of a village or even a house.

FLOATING CATHEDRALS OF ICE Icebergs, so characteristic of Antarctica, are therefore not made from seawater. They are made of freshwater that began as snow falling on the continent thousands of years ago and have traveled thousands of miles before entering the sea. At any one time there may be up to 300,000 of these floating cathedrals of ice in the Southern Ocean. But what we see from the surface is only part of the story. The greater part of an iceberg lies under the waterline. This acts like a huge keel, which gets caught by surface currents of the ocean. The upper part, which can sore over 328 feet (100 m) into the air, catches the wind like a massive sail. Together, wind and ocean currents and the Earth's own rotation transport the icebergs at speeds of up to 25 miles (40 km) per day.

BLOG:

There in the darkness with the ship's spotlights trained upon it was my first iceberg – it was a breathtaking sight. By the morning we were surrounded by sea ice and at 6:00 am in the half-light the view from the bridge was spectacular. A vast white wilderness interspersed with the most beautiful blue-white icebergs.

Eventually, icebergs the size of towns will break up into a myriad of small lumps of ice.

The life expectancy of an iceberg varies greatly, but some have been tracked by scientists using satellites for more than 13 years. Some come to a standstill in shallow waters, where in places hundreds can accumulate into iceberg graveyards. Wave action and warm ocean waters gradually wear and melt them away, and so an iceberg that starts out the size of a town is gradually reduced to fragments the size of a grand piano, called bergy bits and growlers, and of course eventually they disappear forever.

FROZEN PACK ICE Icebergs are not the only ice to float in the Antarctic. Every year the frozen wastelands of the Antarctic are effectively doubled in size by the freezing over of a large part of the Southern Ocean. This sea ice, or pack ice as it's also known, starts to form in May and June, reaching a massive 7.7 million square miles (20 million sq km) between July and

September, bigger than the South American continent. Most of it will melt by the following February to April, shrinking to about 1.5 million square miles (4 million sq km). So although most of the pack ice lasts for less than a year, some can get to be two or three years old. Pack ice is not a static realm but is a rather turbulent place, and the whole landscape can totally change within a few hours. The ice is constantly being jostled, broken up and pushed together by waves, ocean currents and surface winds. Areas of frozen water can open up then rapidly close again. Ice floes pound into each other, their edges grinding together to throw up ice blocks into large ridges dozens of feet high.

GREASE SLICKS OF ICE Pack ice can form in two ways. The first happens in the unsettled weather of the Antarctic autumn. As cold air cools the salty oceans, seawater falls below its freezing point (28.6°F or -1.89°C for salty water, not 32°F or 0°C as with freshwater), and ice crystals form. Because the ice crystals are less dense than the seawater, they float on the surface where the wind sweeps them into large slicks that look remarkably like oil slicks. Because of this and their thick, souplike consistency they are called grease ice.

After a few more hours of freezing, the ice crystals accumulate to form loosely aggregated disks, known as ice pancakes, 2–4 inches (5–10 cm) wide. These merge to form

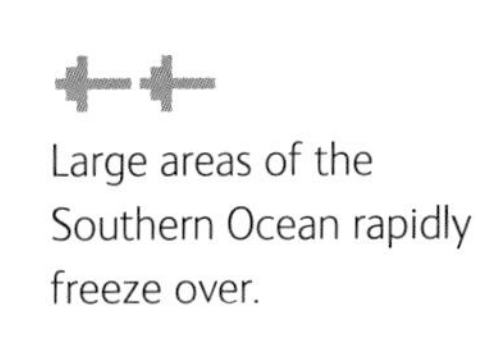

Large areas of the Southern Ocean rapidly freeze over.

Sampling large pancakes of pack ice takes huge skill.

thick super-pancakes several feet across. Wind and waves then raft the pancakes together. By this stage, the ice is strong enough to support a human walking across it, even one with heavy sledges, generators and ice-coring equipment.

FINGER ICE Pack ice that forms under calm conditions is very different. Rather than pancakes, uniformly flat sheets of ice form at the surface. This nilas ice has the consistency of soft ice cream and is prone to finger-rafting, where sheets of ice crack, and pieces dovetail into each other the way fingers interlock. Each piece is known as an ice floe, a slab a few square feet to more than 38 square miles (100 sq km) across. Most Antarctic ice floes are less than 6.5 feet (2 m) thick, although rafting can increase that fivefold. Modern icebreaking ships can break through ice a few feet thick, although in thicker ice even the most powerful of modern vessels get stuck.

So how cold is it really?

The coldest reliably recorded day on the planet was on Antarctica. That day was July 21, 1983, when temperatures plummeted to -128.6°F (-89.2°C) at the Vostok Station, 11,483 feet (3500 m) up on the East Antarctic Ice Sheet. More typical temperatures for the southern winter months at the Vostok Station (May to September) average -85°F (-65°C) and rise to an average of -22°F (-30°C) during summer (December to February). Coastal temperatures are less extreme with an average of between -22 and -4°F (-30 and -20°C) in winter, rising to between 14 and 32°F (-10 and 0°C) in the summer. On the more northerly Antarctic Peninsula the temperatures are on average around 14°F (-10°C) in winter and can be above 50°F (10°C) in the summer.

THE WINDIEST PLACE ON EARTH If the cold doesn't get you, the wind might. Speeds of 186 miles per hour (300 km/h) have been recorded at coastal sites. That's double a hurricane force wind. In fact, Antarctica is often referred to as being the windiest place on Earth. Average wind speeds and snowfall almost have the opposite trend to the temperatures described above. Both are low at the centers of the ice sheets. The air is simply too cold to hold much vapor, and the annual snowfall is about the same as the amount of rainfall that falls on deserts such as the Sahara.

Wind dramatically lowers the apparent temperature because of wind chill.

Huge mountain ranges, where the wind has swept the snow, dominate the Antarctic landscape.

Moving off the ice sheets the katabatic winds (those going down a slope) increase in speed toward the coast. Blizzards are common, causing drifting that can move tons of snow within short periods. There may be so much windblown snow in the air that whiteouts occur, which at the extreme can mean you can't see much beyond the end of your arm. Clearly, working in a whiteout is impossible.

In coastal regions the annual precipitation, mostly as snow, is far greater than that over the ice sheets. Going further north, such as toward the tip of the Antarctic Peninsula, the percentage of precipitation falling as rain increases. However, even on the peninsula there are large climatic differences. On the northwestern side (about 65°S) the climate is a relatively mild maritime climate, whereas at the same latitude on the eastern side the temperatures are about 9°F (5°C) colder due to the effects of the adjacent Weddell Sea.

WIND CHILL The effects of the wind are made more sinister by what's known as the wind chill. Because the speed of the wind increases the rate at which heat is lost from the skin, the temperature experienced is always lower than the actual air temperature. For example, if the air temperature is -13°F (-25°C) and the wind speed 26 feet per second (8 m/s), equivalent to a fresh breeze, the effective temperature is actually -31°F (-35°C). Double the wind speed, and the apparent temperature will be -49°F (-45°C).

BLOG:

We have been endlessly digging snow, which rapidly builds up at the doorways. This is a very important job as we don't want to get trapped in the buildings if there is a fire, one of the biggest risks down here. There was so much snow in the air that if you faced into the wind it was very difficult to breathe.

It's not all snow and ice

Some of the most striking contrasts in this continent of ice and snow are the Antarctic oases, of which the dry valleys close to McMurdo Station in Antarctica's south are the best known examples. Mountains near the coast physically stop the ice sheets, leaving a barren, windswept landscape strewn with boulders and completely free of ice. Dry winds dominate these regions, ensuring the rate of evaporation is greater than any precipitation of snow. The air is so dry that the mummified remains of seals that traveled inland have been discovered there after hundreds of years.

Severe storms are a regular feature of traveling in the waters of the Southern Ocean.

Wind-eroded rocks in the McMurdo dry valleys.

LAKES AND RIVERS A striking feature of the McMurdo dry valleys is that they also contain a series of lakes. Some of these lakes are solid ice from top to bottom, whereas others have an ice cover with highly saline waters beneath. The top 62 feet (19 m) of Lake Vida, for example, is permanently frozen ice capping briny water seven times more salty than seawater. The bacteria living there have been isolated for thousands of years and are now becoming a focus of much research. There are other saline lakes in the Vestfold Hills, such as Ace Lake that also has highly concentrated saline waters. However, on this lake the overlying ice can melt away between January and March. At 77°S, Don Juan Pond, a shallow lake that wouldn't even cover your knees, remains at least partially unfrozen, even when temperatures fall to -58°F (-50°C). It's the saltiest body of water in Antarctica, if not the world.

Despite all the ice, glaciers and lakes, there is only one real river in Antarctica, the Onyx River. This 25-mile (40 km) long meltwater river flows from the Wright Lower Glacier to Lake Vanda in the Wright Valley. Naturally, it only flows in summer and in some years the flow is so low it doesn't actually get as far as the lake. Other localized runoffs from glacier melting spring up throughout the summer, but these are best described as small temporary creeks, as opposed to rivers.

THE PENINSULA The Antarctic Peninsula extends from the northwestern side of the continent, pointing up toward South America. It has the mildest climate of the Antarctic and is the most accessible. Consequently, this is where you'll find most of the research bases as well as most of the tourists. A journey down the western side of the peninsula is spectacular, where huge mountains interspersed by sweeping glaciers form an unforgettable backdrop. Because of the milder climate, and greater proportion of ice-free coasts, the seals and penguins like it, too. There are even patches of green in the summer, with mosses and even higher plants sheltering as far south as 68°S below the Antarctic Circle.

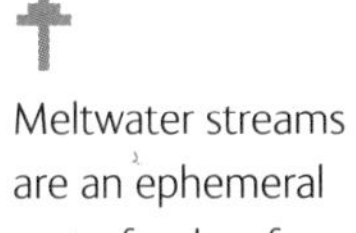

Meltwater streams are an ephemeral part of only a few Antarctic locations.

The mountainous Antarctica Peninsula is ice free in places.

BLOG:

IT'S HARD TO BELIEVE THAT THE LAST TIME WE SAW THE SUN WAS RIGHT BACK IN APRIL. Since then we've all been adjusting in our various ways to living in 24-hour darkness. This long winter night will continue until the first sunrise on August 19th, but thankfully next Wednesday, on Midwinter's Day, we will have reached halfway.

• • •

Now to work

Arriving

FINALLY, AFTER ALL THE PLANNING, medical checks and packing, you're here. You'll be excited you made it, that your dreams of working here and your research aspirations lie right in front of you. But at the same time, you'll feel a sense of isolation. The goodbyes with family and friends linger, and the prospect of long-term separation from everything and everyone you have at home can be tough. Fortunately, there is a huge amount of work to distract you, so you're quickly swept up by the hustle and bustle of settling in and getting down to it. You have to learn the routines for day-to-day living, familiarize yourself with safety issues and, most importantly, get to know your colleagues and learn to work as part of a team.

Only ski-equipped planes can land on the Amundsen-Scott South Pole Station, as it sits on 9350 feet (2836 m) of snow and ice.

SUMMER TEAMS OR OVERWINTERING Most people experience the Antarctic for short stints and generally only between September and March. At many of the stations the population of people staying can quadruple within a few hours of the first aircraft or ship visit after the winter. Imagine the shock of spending the winter with just 10 or 20 people, and then all of a sudden having 100 new faces to deal with.

Increasingly, aircraft is used to transport scientists and equipment onto Antarctica.

Of all the researchers in all the bases dotted around Antarctica, the most hard-core are the overwinterers. They are the ones who have signed up to stay a complete year, sometimes two. They have to survive harsh Antarctic conditions, struggling with the physical and emotional extremes winter brings. A mixture of scientists, medical staff, field guides, cooks and others, overwinterers have the necessary practical skills to keep the stations ticking over during the coldest months. Everything from maintaining the plumbing and power supply through to fixing sophisticated scientific equipment, they do it. The population of Antarctica drastically reduces when winter hits. The American McMurdo Station for example, the largest base on Antarctica, has a summer population of more than 1000 people but is reduced to 200 over winter. In smaller bases, that can be as claustrophobic as 10 or 20 people. Major infrastructure work is left until the new summer teams arrive, when the experienced overwinterers can show them the ropes.

BLOG:

The landscape is just huge, open and intensely bright and clear. The weather has been great with 24-hour daylight but the sun is already noticeably lower on the horizon and every day it gets colder. There were still quite a lot of staff and scientists here when we arrived ... But they have gradually flown out, the supply ships have gone and there are now only a couple more flights before it's just the 14 of us left. 

The Rothera Base on the Antarctic Peninsula.

BLOG:

In Mac Town there is an ice pier. Instead of using concrete, each year they make one out of a big block of ice. They just spray it and spray it to build up the block and put gravel on top so it's not slippery.

Although never luxurious, on the whole living and working conditions are pretty comfortable. Space is often at a premium, so be prepared to share a small bedroom. You might get one to yourself if you overwinter, but you'll probably still have to share a bathroom.

YOUR NEW HOME So what will your new home be like? Antarctic bases vary considerably, from small summer huts to McMurdo Station, which is like a small town. Most are built on rocky coastal areas, but some stand on ice and create a challenge for engineers. Many older stations have been buried under snow and ice, some even crushed. So modern builds are a series of buildings, or pods, raised on stilts several feet up, high above drifting snow. The platforms can be winched up if snow builds up underneath. You can even have buildings on massive skis. Too much snow? Just slide the whole building to safety, although this is not as easy as it sounds with a 55-ton (50-metric-ton) building.

You'll usually find the bedrooms, medical facilities (including operating room) and kitchens at the center of the base. There are separate buildings for workshops, laboratories, food stores, waste collection and water-treatment plants, plus radio rooms and, increasingly, IT facilities. There are the latest technologies for insulation, heating and waste treatment, while fuel depots are kept away from the main building complex for safety. There also have to be provisions for getting on and off the continent, so most stations have helicopter pads. Some also have runways, which if built on ice means the planes need runners instead of wheels.

FLOATING BASES Not all bases are on land. Research ships are effectively floating Antarctic stations, where modern icebreaking vessels operate deep within the pack ice most of the year. With about 50 crew and up to 50 visiting scientists, trips last between one and three months. They launch from South Africa, South America, the Falkland Islands, Tasmania or New Zealand, and may take several weeks to reach the expedition region. Seasickness on the high seas of the Southern Ocean is a major early challenge they face.

The teams on board include a mix of international scientists bringing together a host of skills and analytical tools to answer common questions. The stresses of being away are more intense than on land as you're confined to this floating home for several months with the same faces (plus the same jokes and at times the same annoying habits). The overwinterers would understand. Sampling work happens around the clock and the working day is long, but there is generally a bar, and most ships have a small gym onboard to help you relax. On one research vessel there is even a small swimming pool.

A huge amount of equipment is needed even for the most simple of field camps.

Unloading equipment and vehicles needed for a winter on an Antarctic land base.

BLOG:

Talking of food, it was great to see some fresh lettuce, tomatoes and even basil being tossed into a Greek salad this week. The discovery of such freshies is a big deal, and good for base morale as we haven't seen any for weeks.

... A gangway is put out on the port side to let the 20 or so ice researchers loose onto an ice floe. Small sledges, cool boxes, ice corers and even a canoe filled with electronics are dragged onto the ice for a furious 2 to 5 hours of digging snow pits, coring ice and sucking up water from the underside of the ice ... An army of shadowy figures sample snow and ice lit up by the beams of the ship's powerful spotlights.

FOOD FOR THOUGHT At the heart of your Antarctic home is the dining room. Here you'll meet up with everyone else on the base, share your stories and, of course, eat whatever the cooks have prepared. The cook is one of the most important people around. With a limit to how long fresh vegetables and dairy products can last, they have to use their imagination after the last supplies of the season are delivered, or after several weeks at sea, and make the most of the tinned, frozen and packaged ingredients that remain. In summer, things can get quite busy with different sittings for meals. The dining room is normally the largest communal space and can be the setting for social events, briefings or to watch a film and so on.

Left to right:

Learning to cook in the cold is a vital skill.

It takes considerable ingenuity to sample icebergs.

Teamwork is needed to safely transfer equipment and people between ice floes.

TEAMWORK The key to surviving Antarctica is your ability to work as a team, helping each other keep the whole operation working smoothly. At most stations, teams contain both men and women – before the 1990s they were mostly men. Scientists will have to all pitch in with helping out, such as cleaning communal areas, washing up, sorting the garbage, cooking on the cook's day off and maintaining vehicles and equipment. The nominated base commander is not only responsible for the well being of the team, but also has to have an understanding of all the research projects going on throughout the year. They have to ensure stores of food, fuel, vehicles and medical supplies are sufficient, and that supplies are ordered in plenty of time for the following year.

Research vessels can form a floating base for around 100 people.

Of course, not all the work happens at the stations themselves. For much of the summer there is a continuous coming and going of teams working up to several hundred miles away. Each team needs to be prepared for any kind of weather and is led by very

Helicopters are vital for deploying scientists, sleds and snowmobiles.

experienced mountain or glacier guides. They need helicopters, small aircraft or snow vehicles to get where they're going, along with sufficient food and the right equipment. It's a mammoth task, and with new groups of researchers arriving all the time, all with their own special needs, it requires a flexible attitude.

HOLDING ON TO HISTORY While modern huts and bases are decked out with the latest technology, under international agreement some early huts are maintained for posterity – those that aren't have to be removed from Antarctica in their entirety. The Cape Royds hut on Antarctica's Ross Island was built for explorer Ernest Shackleton's expedition from 1907–1909. A basic build with space to fit 15 in one room, it still has cans of food, equipment and clothing left behind by Shackleton and his team, who had to turn back just 97 miles (156 km) from the pole. It is now listed on the World Monuments Watch List of 100 Most Endangered Sites in the world. Conservators are helping save cans of food, equipment and clothing from the expedition in conditions that are often cold and rather smelly.

YOU NEED TIME OUT Whether on a base or on a research vessel, breaking up the routine becomes increasingly important the longer you're away from home. Mealtimes are brilliant for this, but so is keeping up a fitness regime, fitting in time to read, listening to music (what did we do before iPods and MP3 players?), watching DVDs, painting, taking photographs

or whatever else you do to relax. One particularly ambitious person made a beautiful wooden canoe, and the inaugural launch was in the middle of the Weddell Sea. Every imaginable sport has been played in some form or another on the Antarctic, even croquet and cricket. Marathons have been run, and of course skiing and mountaineering are great diversions. Christmas, New Year, Thanksgiving, Midsummer's Night, national holidays and birthdays are all generally celebrated in style. In fact, any excuse is used for a party. Often presents are homemade, taxing creative skills, especially with the limited resources available. Everyone celebrates the halfway point of any expedition or scientific cruise – from then on you're on the way home.

Fortunately, satellites mean that daily communication with the outside world is easy.

It is not all hard work – ice stumps and a shovel make for an impromptu game of cricket.

E-MAILS, BLOGS AND TELEPHONES A daily exchange of e-mails, blogs and webcams are available from many of the bases and ships. This is a tremendous boost to relaying scientific data back from the field to home. Weather updates and sophisticated scientific information stream from automated measuring stations 24 hours a day, all year round. Even telephone calls are easy to make, though hearing the voices of friends and family can be both comforting and difficult. It can bring the feeling of separation too close for comfort. Despite the hard work, good food, parties and the beauty of it all, everyone needs time on their own. Conditions are cramped, especially during bad weather when you can be locked up for days on end frustrated by not being able to work. Boredom can be a constant threat during these times.

BLOG:

Today, I just haven't been able to get completely warm or comfortable in the lab despite wearing my thermal longs, heavy socks and a beanie. So it comes as no surprise that the temperature recorded this morning was -58°F (-50°C), despite there being little wind.

Challenges

The key to surviving the cold is "layering." Covering up every part of your body with multiple layers of clothing is essential. Goggles, hats and gloves are all vital parts of the kit, as well as very well-insulated boots. Your aim is to stop your body's core temperature (around 98.6°F or 37°C) cooling to the extent that hypothermia starts. Even a one- or two-degree drop can bring on the early stages, such as disorientation and lack of logical reasoning. This can be a disaster if help is not at hand.

FROSTBITE Gloves, face masks and fur-edged hoods are vital to prevent dreaded frostbite. Your body closes off the blood supply to extremities such as your fingers, toes and nose when it is cold. A combination of freezing temperatures and low blood supply can lead to tissue

It is essential to keep your extremities, such as your nose, covered.

damage, especially if the skin is exposed. The first signs of frostbite are a hardening, whitening and numbing of the skin, often accompanied by a burning sensation. At this stage, you need to warm up the skin as quickly as possible. But left untreated, things get worse. The affected skin becomes very painful and starts to swell. It then starts to darken, and after a few hours will eventually turn black as if burnt. If the nerves and blood vessels have been severely damaged, gangrene may follow, and the affected part may need to be amputated. One of the major problems with even mild frostbite is that once an area of skin has been affected, it is more susceptible to frostbite in the future.

DRYING OUT You might be surrounded by frozen water, but dehydration is a common problem in Antarctica, especially when working in the extreme dry of the inland deserts and valleys. Combine the dry air with cold temperatures and working while being weighed down with bulky clothing, and you see how easy it is to lose water. The best check is your urine. If it's dark and concentrated, you're dehydrated. If not corrected quickly by drinking plenty of fluids, you can start to feel light-headed and nauseous and your thinking becomes muddled.

IT'S EASY TO GET SUNBURNT Although temperatures are freezing, summer brings long days and bright sunshine. It doesn't take long for the glare on the snow to crisp up any unprotected skin. Much of the harmful ultraviolet radiation is usually absorbed by the ozone layer, a layer in the Earth's stratosphere, some 6–31 miles (10–50 km) up. But ozone depletion over the Antarctic has been well documented in the past 20 years. From September to December each year, the reductions in ozone concentration cause ozone holes in the layer, which let through more ultraviolet radiation over large parts of Antarctica. The sunblock you packed will protect any exposed skin and a very good pair of sunglasses will prevent damage to your eyes.

Even in the summer overalls, gloves and sunglasses are vital for work.

SLEEP PROBLEMS While hot nights keep you awake during the summer at home, summer in Antarctica is plagued by 24-hour daylight. It's good for when you're working late but plays havoc with your sleep. A good set of curtains or black-out material is useful for darkening the bedroom. But winter has the opposite problem, with 24-hour darkness or at best gloom. It's important to keep the windows uncovered to maximize the little light you're exposed to. The lack of daylight strongly influences your natural body rhythm of day and night, which can affect your sleep as well as your eating. Some people use special artificial lighting to help stimulate their metabolism.

SELF-CLEANING HAIR AND SNOW BATHS On the bases and on ships, water is in good supply (there is lots of snow to melt) and washing is not a problem. At field camps this may be more of an issue, although there is nothing more refreshing than a snow bath! We're obsessed with cleaning ourselves at home, but after a few days of not washing your body will start to self-clean. Even greasy hair. Fortunately, at low temperatures, bacteria growth is slow. Outside in the cold your sense of smell goes too, so you don't pick up offensive body odors from yourself or your companions. That changes once back in the warmth of the base, where smell is quickly regained and you realize you urgently need a long shower.

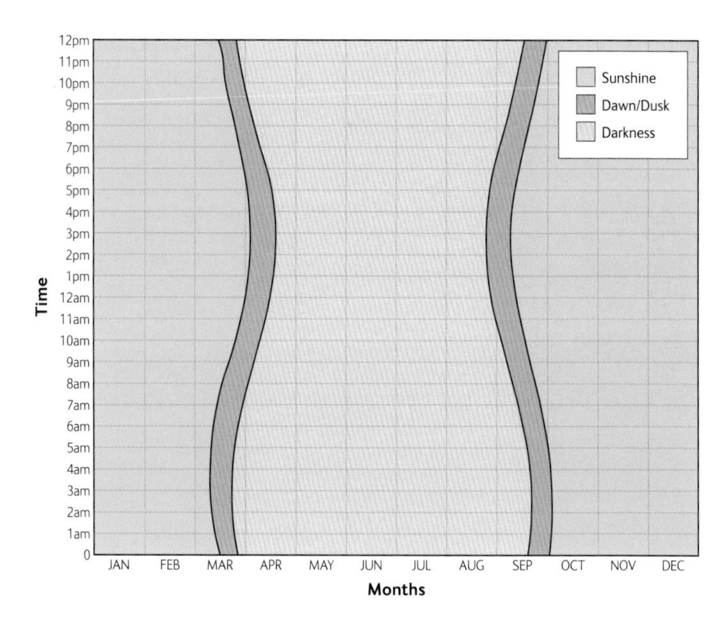

Sleeping at a field camp is not just cold, but also difficult with 24-hour daylight.

A large part of the Antarctic year is dark, whereas in summer the sun never sets.

The glow of
the midnight sun.

BLOG:

SOON AFTER DIGGING THE SNOW PITS ...

a curious emperor penguin slid and waddled between the groups taking a long hard look at what was going on. A few hours later it was joined by a friend, and when we left the floe there were five of these splendid looking birds basking in the sun. By this time they had turned into rather aloof stars who simply know they are wonderful.

Viruses to whales

The tiniest life

FREEZING TEMPERATURES, WIND CHILL, dehydration, ice, low light and ultraviolet radiation – it's amazing anything survives in the Antarctic. But so much does. The oceans around it, the coastlines and even the dry valleys are home to unique and thriving fauna and flora. Everything from viruses to blue whales is found here, each adapted to these extreme conditions. The story begins under the microscope, since microscopic organisms are the founding blocks of the food chain and are more important than largely appreciated. To emphasize the point, the weight of all the viable bacteria in the Southern Ocean is more than that of all the whales.

ANCIENT ICE Bacteria have been extracted from glacial ice over 20,000 years old from sites as diverse as Greenland, the Antarctic and high Tibetan Plateau. In the Vostok ice core, bacteria have been retrieved from ice 2 miles (3.5 km) from the surface and just 492 feet (150 m) above the subglacial Lake Vostok. The ice at this point of the core is actually frozen lake water stuck to the bottom of the overlying ice sheet, so scientists are predicting that viable bacteria will be present within the ancient lake waters further down. Many of the species recovered from these ancient ice samples are spore-forming species, which are very resilient resting stages of the bacteria that enable them to survive harsh conditions. Clearly they have survived very well for many thousands of years. These ancient bacteria may give us clues as to how life-forms survive on the ice of extraterrestrial systems such as on Mars or on Europa, the icy moon of Jupiter.

COLORED SNOW Not all the microbes on Antarctica are ancient, and some can have quite startling effects on their surroundings. Against the vast white expanse, huge patches of red, green and yellow can appear on the snow surface, caused by microscopic algae. The various colors come from the combinations of pigments produced in the algal cells for photosynthesis and to protect against ultraviolet radiation. To grow, the cells need water, light and a source of nutrients such as nitrogen and phosphorus. They get water from the melting snow. The tiny amounts of nitrogen and phosphorus they need are also trapped in snow. Where there are animals, especially birds, their feces and guano are a rich supply of nutrients, supporting a richer growth of algae.

The footprints of Adélie penguins are highlighted by snow algae that move through the snow.

EXTREMELY TOUGH SNOW BACTERIA Bacteria have been found in snow as far south as the south pole. Among the toughest snow bacteria is the genus *Deinococcus*. They can survive extreme drought, lack of nutrients, blasts of high ultraviolet radiation and even extreme ionizing radiation levels. Recent studies have shown that there can be an extremely thin layer of liquid water on snow and ice crystals down to -36°F (-38°C). This layer is about the thickness of a bacteria cell and may provide just enough water for these bacteria to survive such temperatures.

WATER IN UNLIKELY PLACES When rock debris such as loose rubble or dust blows across snow and ice, it can initiate a unique type of habitat for microbial life. Known as cryoconite, the tiny rock particles absorb heat while their surroundings remain frozen. The warmed dust melts the snow underneath it, forming a small hole (cryoconite hole), typically several inches deep. The meltwater collects at the bottom of the hole and is colonized by viruses, bacteria and photosynthetic algae. Organisms such as protozoans, a group of single-celled organisms, are also found in these holes, feeding on the bacteria and algae. Several holes can melt together to form larger pools. Melt ponds a few square feet across and several feet deep are also common on glacier surfaces in summer.

The freezing and thawing of groundwater often sorts rocks and stones into defined polygon patterns in regions of stone rubble fields. Hypoliths are a group of microorganisms – mostly cyanobacteria and single-celled algae – that grow on the underside of the stones and rocks, where they photosynthesize in light far less than 0.1% of the sunlight reaching the ground. These organisms are certainly growing, even thriving, in an extreme environment, where temperatures sink below -22°F (-30°C), water is minimal and light conditions are reduced to virtually nothing. This is thought to be important as the base of a very short food web, which is basically

Getting ready to sample life under the ice.

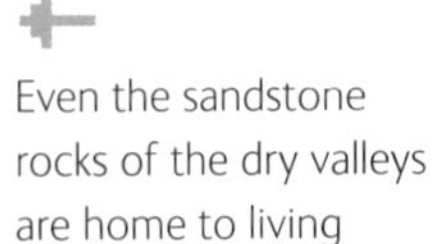

Even the sandstone rocks of the dry valleys are home to living microbial assemblages.

just nematodes (tiny threadlike worms) and protozoans that are feeding on the very slow-growing hypoliths. However, they are estimated to be just as productive as other photosynthetic organisms, such as the lichens that sparsely inhabit such regions.

It's not just under stones that microbial activity flourishes. The quartz sandstones of the dry valleys and other regions can be home to microbes, including fungi, algae and bacteria. These organisms live in small cracks and even between the grains of porous stones and rocks. Quartz and sandstones are just about transparent to light, at least the first fractions of an inch from the surface. The organisms are highly adapted to photosynthesizing and growing in this low light, as well as using the trace amounts of nutrients available within the rocks. Naturally, moisture is still essential for their survival, but the porous nature of the rocks does mean that, once wet, they retain tiny pockets of water between the grains that the organisms can use.

INLAND LAKES The lakes of the dry valleys to the south and the Vestfold Hills to the east have received a lot of scientific interest, especially in recent years. The plankton in these salty lakes are a mixture of bacteria and algae, adapted to low temperatures, low light and sparse nutrients. Some species slightly higher up the food chain but still single celled, such as protozoa, can switch between photosynthesizing when light levels are high in summer to ingesting bacteria and algae to survive the gloomy winter months. Some lake inhabitants gather just under the ice to soak up the maximum amount of light, while other low-light-adapted species seek out warmer waters dozens of feet down. Each also has a specific tolerance to salt, which governs where in the lake they live. The further down you go, the saltier it gets.

In recent years there are increasing reports of archaea being found in permanently frozen Antarctic lakes. Archaea look like bacteria, but taxonomists have separated them into a separate kingdom from the true bacteria (eubacteria). They are an interesting group, since they are typically found in hostile habitats such as thermal waters, acidic waters, on and within crystals of salt and even in deep cracks within the Earth's crust. It is not surprising, therefore, that they are found in these extreme Antarctic habitats.

Scientists sample the biology that lives within lake water.

LIFE INSIDE THE PACK ICE

Underneath a pristine, white snow cover of pack ice, the ice is often a light brown to rich coffee color, caused by a thriving multitude of tiny (mostly microscopic) organisms — viruses, bacteria and algae. They include many of the smaller

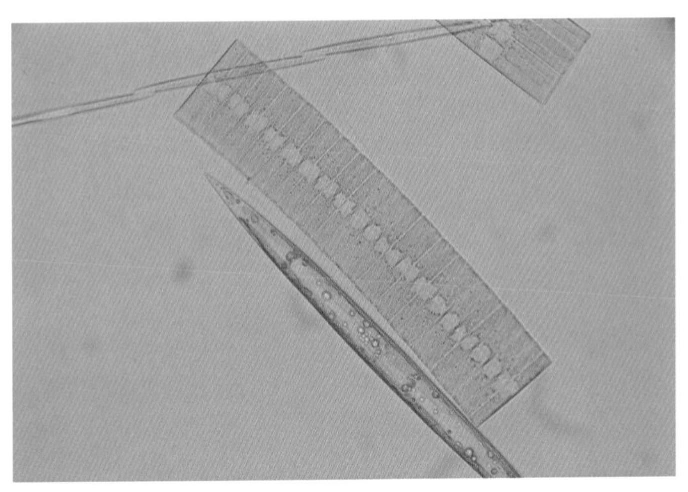

Microscopic algae reach such concentrations in pack ice that they turn the ice brown. They are the food for many zooplankton, including the Antarctic krill.

constituents of the plankton of the open ocean, which is in fact where they originate. As ice forms, plankton organisms stick to, or are caught between, ice crystals as they rise through the water. The trapped pack-ice organisms face considerable changes in their immediate physical and chemical surroundings as ice grows – ice crystals are pure water, and the minerals that make seawater salty are concentrated in a liquid brine. Compared to ice formed from freshwater, pack ice is not solid but is laced by an intricate network of channels and pores filled with the brine. It is within this salty labyrinth, which looks like a liquid-filled Swiss cheese or sponge, that the organisms live. The most conspicuous organisms in pack ice are pennate diatoms (single-celled photosynthetic microalgae), which reach such concentrations that their photosynthetic pigments discolor the ice brown. Bacteria are also very numerous in pack-ice communities and are mobile and active between 14 and -4°F (-10 and -20°C), the temperature of most domestic freezers. Protozoans, turbellarians (flatworms) and small crustaceans also accumulate in pack ice at concentrations ranging from hundreds to thousands of individuals per liter (about 4 cups) of water. These organisms feed on the dense growths of bacteria and diatoms within the brine channels.

Not only do these ice organisms have to survive freezing temperatures, but they have to be able to cope with the high salt concentrations in the brines. Diatoms and bacteria do this by producing a range of chemicals inside the cell that act as antifreezes, and also in the osmotic control of water inside the cell. They also secrete a range of mucus-like substances in which they encapsulate themselves to provide protection from the most extreme conditions.

A TOUGH LIFE FOR VEGETATION Unlike the hardy microorganisms, most of Antarctica's plants died out about 55 million years ago. Subantarctic islands, such as South Georgia in the north, are still lush green places, at least in summer, but there are no trees or shrubs on the continent. Vegetation is restricted to a scattering of 100 or so species of moss, 25 species of liverworts and up to 300 lichens. Some mosses are active below 27.5°F (-2.5°C), and lichens have been known to photosynthesize at temperatures as low as 1.4°F (-17°C). With generally low levels of moisture, poor and freezing soils, lack of insect pollinators and long periods of twilight and darkness on Antarctica, it's easy to see why vegetation is so sparse. Only two higher plants flower, a grass, *Deschampsia antarctica*, and a pearlwort, *Colobanthus quitensis*. Both of these

Lichens are some of the hardiest organisms, withstanding drying out in combination with freezing temperatures.

only grow as far south as 68°S, whereas mosses and lichens have been found much further down, at 84°S and 86°S respectively.

Most of the vegetation grows on the western side of the Antarctic Peninsula, in its maritime climate. Here the soils are better, and the higher precipitation provides the necessary moisture. The penguin colonies provide a rich source of fertilizer through their guano. Vegetation is never thick or lush – rather patches grow in sheltered clumps of soil or in rock crevices. But it's growth nonetheless. Lichens are the most resilient forms of vegetation, being able to shut their metabolism down in winter, where they stay effectively freeze-dried until rehydrating in the moist summer air. Their growth is slow, but if undisturbed some species can survive for thousands of years.

IT'S A DIFFERENT STORY UNDERWATER Several seaweeds can tolerate freezing at -4°F (-20°C), and some have even been shown to survive -76°F (-60°C). Few seaweed species can exist on the intertidal zone of Antarctic coastlines, not due to cold but because of the rhythmic scraping by the pack ice as it floats and crashes against the rocks with the tide and waves. Only a few fast-growing, small seaweed species and tough, encrusting lichens survive such conditions.

Below the low-tide mark, however, ice scour effects are less pronounced, and about 100 seaweed species grow down to depths of 65 to 164 feet (20 to 50 m). Large brown seaweeds are so suited to this environment that they can form dense kelp forests. Some species, including the red *Iridaea cordata*, grow at latitudes of 77°S, where they grow under pack ice for up to ten months of the year. They and other seaweeds that experience long periods of seasonal ice cover have specialized photosynthetic pigments and metabolisms to exploit low light levels, enabling them to grow virtually in the dark.

The red seaweed *Palmaria decipiens* gets a head start on its growth in the winter months when overlying pack ice blocks out the light entirely. Instead of using light, it mobilizes starches built up in the previous year's growth and forms new growth. The new tissues are ready to begin photosynthesis as soon as the spring light appears. When the ice clears, it grows as quickly as species in much warmer waters.

BLOG:

... Our engineer will do a check of the pump filters. He often finds an array of sea creatures including spiders, lice, worms and various fish. The lice are particularly amazing and are surprisingly large ... with long, elongated legs. They have a dark colored back, but the underside of their bodies is creamy white in color and translucent.
I haven't seen anything quite like them before. They are kind of creepy looking and a little out of this world. Makes you wonder what else is lurking about under the ice.

Invertebrates – no backbone

There are no terrestrial vertebrates on Antarctica – penguins and seals do not live on the land, they just use it as a resting and breeding place – but there are some invertebrates, those animals without a backbone. Like the plants, numbers are low and poorly distributed. They are also mostly very small and include animals such as nematode worms, springtails and mites. It's not so much the freezing temperatures that limit their activity, but rather the threat of dehydration. Some produce antifreezes that enable them to combat the threat, enabling them to cool down to between -4 and -58°F (-20 and -50°C) without freezing. The mite, *Nanorchestes antarcticus*, is considered to be the most southerly occurring animal, at 85°/35'S. It not only tolerates temperatures as low as -40°F (-40°C), but remarkably it can also survive very warm temperatures of 99°F (37°C).

LIFE ON LAND Species of the small, segmented group called tardigrades (water bears) and the multicellular rotifers have devised a very effective survival strategy. These organisms live in the soil and plant debris, as well as in the bases of lichens and within cushions of moss. When external conditions get dangerous, say temperatures plummet or there is high radiation, they undergo a state of cryptobiosis where they halt their metabolism and reduce their water content down to as little as 1%. Essentially, they go into suspended animation. It's thought tardigrades might be able to stay this way for hundreds of years.

Springtails, at 0.07 inches (2 mm) long, are one of the biggest land-living animals on the continent. They can be found in swarms around penguin colonies, eating mainly dead vegetation and fungi. Soil-living nematodes are also numerous and are vital for the breakdown of organic matter and the resulting release of nitrogen and phosphorus into the ground. They feed on bacteria, fungi and algae, and can also enter cryptobiotic states. Not to be forgotten, of course, are the nematodes, fleas and lice that live within the fur of animals, between the feathers of birds and among the baleen plates of whales. These are surely among the warmest places in Antarctica for invertebrates to live. However, the invertebrates that have it best are the numerous parasitic organisms that live inside the mammals and birds.

INVERTEBRATES UNDER THE SEA In contrast to the few invertebrates on land, the sea is heaving with invertebrate life. Many are sessile suspension feeders such as sponges, jellyfish and echinoderms. They live on food particles in the water, from bacteria to small crustaceans. As with the cold, deep sea some species of Antarctic mollusks and crustaceans can grow much larger than their counterparts in the shallow waters of warmer regions. Antarctic sea spiders, for example, at 16 inches (40 cm) across — bigger than a large dinner plate — are 100 times the size of the common European sea spider. The woodlice-like isopods, such as *Glyptonotus antarcticus*, found throughout Antarctic waters, grow up to 8 inches (20 cm) long. Isopods in other parts of the world may reach a paltry couple of inches. Other "giants" include sponges 6.5–13 feet (2–4 m) tall and ribbon worms 10 feet (3 m) long.

A research diver reaches out to a jellyfish. He can only stay under the pack ice for about 40 minutes.

So what causes this gigantism? It's thought to be a combination of factors. Low temperatures slow down the animals' metabolic rates. Consequently, slow growth rates enable them to live longer. They may grow more slowly, but because Antarctic animals live longer, they grow bigger in the long run than similar species from warmer waters. Of the sponges, many are showstoppers. Some, such as the white volcano sponge, *Rosella nuda*, can grow as tall as an adult human. And the Antarctic lollypop sponge, *Stylocordyla borealis*, can live up to 150 years. The largest of the *Rossellid* sponges, about 6.5 feet (2 m) high, is an estimated 10,000 years old, which if true would make them the oldest living colonial organisms on Earth.

Shrimplike krill are the food for whales, seals, birds and squid.

THE ANTARCTIC KRILL One tiny invertebrate is food for many of the very much larger Antarctic mammals and birds – krill. There are estimated to be about 1.7 billion tons (1.5 billion metric tons) of these shrimplike crustaceans, *Euphausia superba*, in the Southern Ocean, and weight-wise, there are three times more krill than humans on Earth. In terms of biomass, it is one of the most successful species known; between 168 and 336 million tons (150 and 300 million metric tons) of them sustain squid, seals, birds and whales each year. So crucial is their role, they are often referred to as a keystone species in the Southern Ocean's food web. Krill grow to about 2.5 inches (6 cm) long and feed voraciously on plankton in the water. A group of krill moving through the water is rather like a swarm of locusts – they devour anything in their way. They form swarms so dense that if a ship passes through them, the bow wave flushes a spectacular blood red.

They, too, are one of Antarctica's survivors. In winter when there is little food, adult krill can withstand extended periods of starvation, living from lipid reserves and even shrinking – unusual for a crustacean with a hard exoskeleton. In laboratory experiments, krill have been shown to survive up to 200 days like this. In more plentiful times, they come to the surface to feed at night, swimming the deeper waters during the day. In turn, they are followed by bigger animals such as penguins, seals and fish. But in winter some krill populations use cracks and crevices in the ice to hide from these predators. In years when pack ice cover is prolonged, krill populations swell, and vice versa in poor pack-ice years.

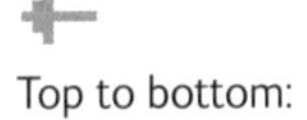

Top to bottom:

An underwater camera connected to onshore equipment allows scientists to study life all year round.

A translucent jellyfish floats through the icy water.

Sea stars rest at the bottom of Mcmurdo Sound.

As we traveled through the ice, it is amazing to watch the thousands of krill (shrimplike organisms) washed up onto the overturned ice floes. They are the obvious food for the large number of seabirds that congregated in the region.

YOUNG KRILL CARE The larval stages and juvenile krill do not have enough reserves to starve for any considerable time. So in winter they survive by feeding on algae and bacteria on the underside of ice floes. They scrape the food from the ice in such an efficient way that larval krill can clear an ice surface of all attached microbes at a rate of one-fifth of a square inch (1.5 sq cm) every second.

Fish and squid

The Antarctic waters make up about 10% of the world's oceans, but only hold about 1% of the world's fish species. However, the majority of fish found here are only found here; they are what's called endemic. Unusually, most spend their time in deep waters, more than 656 feet (200 m) from the surface. This is probably to avoid the pack ice and to stay where temperatures are more stable. Few have swim bladders, used for buoyancy in other species, and about half live on or close to the seafloor. Despite not having swim bladders, many of the species have highly cartilaginous skeletons (thus reducing the weight) and fat deposits that help them maintain their buoyancy.

FISH HIDING IN PACK ICE One fish to swim in surface waters is the broadhead, *Pagothenia borchgrevinki*, which takes advantage of the algae growing on the underside of pack ice. It's even been seen clinging to the underside of the ice when resting and uses crevices and holes to hide in. Young stages of the Antarctic silverfish, *Pleuragramma antarcticum*, have also been seen feeding on the underside of ice floes as have, on a few occasions, young stages of the giant Antarctic toothfish, *Dissostichus mawsoni*. These two species normally live in deeper waters but migrate to the underside of the ice to feed extensively on krill.

ICE FISH Hemoglobin and red blood cells are important in most animals for carrying oxygen around their bodies. However, oxygen is much more soluble in the cold Antarctic waters than elsewhere. Also, Antarctic fish generally have very low metabolisms, cutting down the need for high oxygen content to be carried in the blood. As a result, many Antarctic fish have low red blood cell counts and hemoglobin levels compared to fish from warmer waters. The 15 species of Antarctic ice fish (*Channichthyidae*) have no hemoglobin in their blood at all.

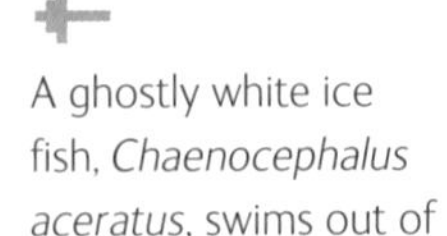

A ghostly white ice fish, *Chaenocephalus aceratus*, swims out of the way of a diver.

As hemoglobin is what makes red blood cells red, these fish are a ghostly white, hence their name. Ice fish also have a much greater volume of blood and larger hearts compared with similar-sized fish elsewhere, enabling them to transport oxygen without having to produce specialized blood pigments, which requires energy.

ANTIFREEZE FOR FISH BLOOD The blood of many fish freezes at about 30.5°F (-0.8°C), which is 1.8°F (1°C) higher than the surrounding seawater. Clearly, few fish would survive if that were the case in practice. Instead, special antifreezes lower the freezing point of the blood close to the freezing point of seawater. There are a wide variety of these antifreeze glycoproteins (AFGPs), and the broadhead produces an impressive eight of them. The best-studied fish for AFGPs is the largest of the Antarctic fish, the giant toothfish, which at up to 310 pounds (140 kg) weighs about the same as two adult humans. It can live for 45 years. Because its heart beats only once every six seconds, it's also used in medical studies to see how hearts behave during certain cardiac treatments.

ANTARCTIC SQUID There are thought to be about 112 million tons (100 million metric tons) of squid in Antarctic waters. Not much is known about these enigmatic creatures, which frequent deeper waters and only surface at night. But we do know from finding squid parts in the stomachs of whale and seal species that they are a crucial part of survival for many Antarctic animals. While Antarctic fish provide about 17 million tons (15 million metric tons) of food for birds, seals and whales, squid are thought to provide more than twice that.

FAST GROWING, BUT NOT FOR LONG Squid do not live very long. Even the giant species have a lifespan of only one year. This means they must grow quickly, especially the larger species, and so they eat about 30% or more of their body weight each day. While most animals stop growing in later life, squid continue growing at a constant rate until they die.

There are about 20 Antarctic squid species, from the small *Brachioteuthis* that would fit in your hand to the large *Mesonychoteuthis hamiltoni* that's almost as long as a great white shark. This beast catches prey such as the giant Antarctic toothfish using large hooks, and suckers on its arm and tentacles. In general squid feed on a range of fish and crustaceans, including the ubiquitous krill. They will also eat other squid.

The birds

Although birds do live and breed on the Antarctic continent, there is virtually no food for them on land, so most depend on the Southern Ocean for food. Trips out to sea can mean long journeys to and from nesting sites. As a consequence, some can travel epic distances, living off their fat reserves, if necessary to the point of starvation. Large colonies of breeding birds can gather, and for reference they are some of the smelliest places in Antarctica. You won't forget the stench of a penguin colony in a hurry.

GRACEFUL WADDLERS Which brings us to the most familiar image of Antarctic wildlife – the penguins. Worldwide there are about 17 species of penguin, but few are found south of the Antarctic Circle. It's only the emperor penguin, *Aptenodytes forsteri*, and Adélie penguin, *Pygoscelis adeliae*, that are characteristic of the pack-ice regions. Imagine working on the edge of an ice floe, the ocean floor 2.5 miles (4 km) below. All of a sudden there is a commotion, a spray of water, and an Emperor penguin shoots up from the sea, skidding along on its tummy a few feet away. It's not clear who is more surprised, you or the penguin, as it puts its beak into the snow and levers itself into a standing position. Very soon several others have catapulted themselves from the water onto the snow, nodding and calling to each other. Are they discussing this strange red-suited creature, or perhaps the krill and fish they have just been hunting, or a near miss with a predatory leopard seal? Whatever it is, their cooing is one of the most evocative sounds of Antarctica.

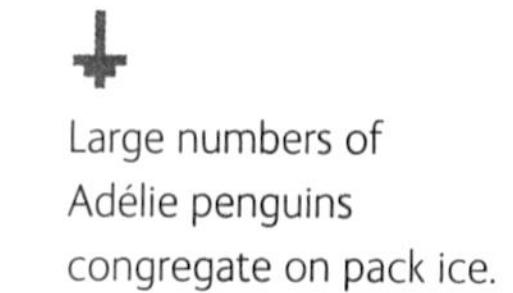

Large numbers of Adélie penguins congregate on pack ice.

BLOG:

They [Adélie penguins] are very inquisitive and will walk up to you if you stand still. They grunt and bark, with a single bark each time. I mimicked one of them and to my surprise it barked back tit for tat much like a cat, had me giggling ... Watched them walk away, which was delightful. As soon as there is a hint of a decline they dive onto their bellies and push along with their feet, very cool.

On land, penguins look quite ungainly, either waddling around or gliding along on their tummies while propelled by their feet. But in the water, they come into their own. Like all penguins, the "wings" of these flightless birds are reduced to flippers, brilliantly effective as paddles. They are amazing swimmers, routinely reaching up to 6 miles (10 km) an hour, and capable of short bursts twice this speed. Adélie penguins dive for several minutes to catch krill and fish in the top 165 feet (50 m) of the ocean, but can go to 575 feet (175 m) if need be. Emperor penguins, when feeding on squid, can dive more than 1805 feet (550 m) below the surface, staying under for more than 20 minutes. It's a magical sight to watch flocks of penguins leaping from the water at great speed and arching together through the air, catching a breath before diving into the water again. This technique of taking quick breaths allows them to maintain their speed in the open water.

BREEDING EMPERORS Emperor penguins have so far been discovered breeding in about 30 to 40 colonies on Antarctica. Adults arrive in March and April, all fattened up. Females lay a single egg, pass it to the male to look after, then slide off to feed, traveling up to 95 miles (150 km) across the pack ice to get to open water. They won't be seen again until their eggs hatch. Then comes the endurance test for the males. They must wait 60 or 70 days in temperatures below -40°F (-40°C) for the chicks to hatch. They can't leave the eggs to go and feed, and they must huddle with other males to keep warm, taking turns to be on the frozen edges of the group. All the while, the egg nestles in the brood pouch a full 248°F (120°C) warmer than outside. To maintain these conditions, the birds have to expend considerable energy, burning up their fat reserves at rates of up to 7 ounces (200 g) per day. Over the incubation they may shed 50% of their body weight this way.

July or August is hatching time. The males rise to the challenge of feeding the emerging chicks despite their dwindling energy reserves. They produce a rich secretion of fat and protein for the chicks to

feed on. Soon after, the females return and take over feeding the young with regurgitated food while the males return to the sea to feed. Imagine having to travel 100 miles (160 km) for your dinner after eating nothing for over three months! It is an epic feat, after which parents take turn to feed the young. Despite every effort, some young starve if the ice is particularly extensive that year. It just takes too long for adults to get to and from the ice edge for food. Those that do survive go for their first swim to hunt for food in December, when the adults finally stop feeding them. The young penguins' insulating, downy covering is gradually shed and replaced with stiff feathers packed closely together. These not only conserve heat by trapping air close to the body, but also form an efficient waterproof layer.

A group of emperor penguins getting ready to have their first swim of the season.

INQUISITIVE ADÉLIES The most curious, and watchable, are the Adélie penguins. Emperor penguins remain aloof and keep their distance, whereas the Adélie penguins take a great interest in what you're doing on the ice. They are the most numerous of the Antarctic

Emperor penguins travel long distances sliding on their fronts.

penguin species, with around two million breeding pairs. Although these small penguins eat fish, the bulk of their diet is krill. In contrast to emperor penguins, Adélies spend the winter at the most northerly edges of the pack ice, up to 435 miles (700 km) away from land. Before the pack ice breaks up in late winter, the adults walk and swim back to the breeding colonies on the main continent. Their round trip, out to winter grounds then back to breed, can be exhaustingly long. Some birds have been tracked by satellites traveling up to 3100 miles (5000 km). Adélie breeding colonies are all in places that are ice free in the summer, when the chicks hatch. That way, the young can feed before traveling north for the winter months. Parents typically travel up to 20 miles (30 km) each day to find food. Some have been tracked for more than 125 miles (200 km).

A young Adélie penguin shedding its juvenile feathers.

White snow petrels are perfectly camouflaged in the snow and ice.

PERFECT CAMOUFLAGE As evocative as the penguins are, they are not the only birds to exploit and survive the Antarctic. Flocks of white snow petrels, *Pagodroma nivea*, twist and turn against the backdrop of a cloudless blue sky. If they land on the snow, you can hardly see them at all, their white plumage providing a perfect camouflage. These are common birds in coastal regions, feeding on fish and crustaceans in the sea. They sit on the edges of ice floes waiting for prey to venture from under the ice floe edge, when they will dive in to catch it. Nesting on rocky ledges or cliff faces, some have even been found more than 112 miles (180 km) from the coast on mountain summits. This is remarkable behavior, considering the adults must fly backward and forward to the pack ice to collect food for their chicks.

GIANTS OF THE AIR Rather larger is the southern giant petrel, *Macronectes giganteus*, with a wingspan of up to 6½ feet (2 m). These huge birds are widely distributed over the Southern Ocean and soar eerily high over the pack ice. Mostly solitary, these scavengers congregate around seal and penguin carcasses, although they have the strength to kill smaller birds for themselves. The most majestic of the giants is, of course, the albatross. The largest species, the wandering and royal albatrosses, stretch their wings to an impressive 11 feet (3.5 m).

A scientist measures the bill length of a giant petrel chick.

They have the largest wingspan of any flying birds in the world and can live more than 60 years. They breed on subantarctic islands, and they breed for life, from the age of about 13, courting with elaborate dances that rival any ballet.

The albatrosses do not look for food in the pack ice, preferring the open waters. You could spend hours on deck, watching one soar majestically above the waves hardly flapping its wings, occasionally dropping into the sea to snatch a squid, fish or large crustacean. Their huge wings make ideal gliders, using the air currents at the surface of the ocean, without the need for wasteful energetic flapping. Their flight is so efficient that they have been recorded traveling up to 5000 miles (8000 km) in a single week in search of food.

Seal survival

Mammals have also adapted to life in the frozen south. You won't find polar bears here, but what you will see are seals, lumbering ungraceful hulks scattered on the ice floes. There are five species common in southern Antarctic waters – the southern elephant, Weddell, crabeater, leopard and Ross. A sixth species, the fur seal, is the most common species in the north. A priority for all seals is conserving heat and having waterproof pelts, essential while swimming. Many lay down vast reserves of fat, or blubber, which make a fantastic insulator and a good reserve of energy. In the lead-up to winter, when food becomes scarce, some seals lay down as much as half their body weight as blubber. The fur seals, as their name suggests, have perfected a different way to stay warm. Instead of producing lots of blubber, they have a double layer of insulating fur. All Antarctic seals have an outer layer, but the fur seals have both a coarse top layer of hair and an inner layer of finer fur that acts as superb insulation.

HOLES IN THE ICE The Weddell seals, *Leptonychotes weddelli*, breed farther south (78°S) than any mammal on Earth. Unlike other Antarctic species, which breed on floating pack ice, the Weddell seals congregate on the coast where the pack ice joins the land or ice sheets.

Weddell seals breed farther south than any other mammal.

Young seals put on weight very quickly, fed on a fat-rich diet of their mothers' milk.

Seals tagged with satellite transmitters can give much data on migration patterns and diving behavior.

Here they give birth in September, when air temperatures are around -4°F (-20°C). What a shock for the pup to leave the cozy womb at 98.6°F (37°C), and the next minute to be on the freezing ice. So mothers quickly feed them a very rich milk, containing 60% fat. The pups quickly double their weight on this fatty diet. Within two weeks they are encouraged to have their first swim, and within 12 weeks they are completely independent.

Mothers and pups access the water through holes in the ice, which the seals keep open by grinding away the edges with their teeth. Unfortunately, the grinding wears their teeth down prematurely, eventually making it difficult for them to feed. Most die young from starvation. Weddell seals are expert hunters. They feed mainly on squid and fish, in particular Antarctic silverfish. They can dive as deep as 2300 feet (700 m), for as long as an hour. And they have two dive patterns, matched to where the fish are at certain times of the day. They use shallow dives, about 330 feet (100 m) down, at night when the fish come up to feed on krill. Deeper dives happen during the day, from 655–1310 feet (200–400 m) down, again probably due to where the fish are.

CRABEATERS DRIFTING ON THE PACK ICE With 30 million crabeater seals, *Lobodon carcinophaga*, in the Antarctic they are thought to be the most numerous seal species on Earth. They spend their entire life drifting with the pack ice, and can travel many thousands of miles a year this way. As ice melts in spring and summer, the populations of crabeater seals move further south with the receding ice edges. They give birth to their pups on the ice, dotted around the ice floes or huddled together in large groups. Like the Weddell seals, crabeater pups are weaned quickly, and their first swim comes after only a few weeks. They synchronize their hunting with local food, hunting mostly at night when the krill rise to the surface waters. Their special interlocking teeth help filter krill out of the water. They can dive down to 1740 feet (530 m), but more usually you'll find them hunting at 65–100 feet (20–30 m), and for short bursts of five or ten minutes at a time.

LEOPARDS OF THE SEA Leopard seals, *Hydrurga leptonyx*, have a reputation as fearsome predators. Look into a leopard seal's mouth, at the impressive array of teeth, and you can see why. Half their diet is krill, but they also eat fish, penguins and crabeater seal pups. It is an awesome sight to watch a leopard seal tossing a young seal or penguin into the air again and again, flaying it with repetitive slaps against the water. Many crabeater seals have telltale wounds from leopard seal attacks. These aggressive and agile hunters swim around ice floe edges or close to penguin colonies, waiting for their prey to enter the water.

THE MYSTERY ROSS SEAL Not much is known about the Ross seal, *Ommatophoca rossii*, a curious animal with almost no neck. They are rare and the smallest of the Antarctic seals. It's thought they feed predominantly on krill, mid-water fish and squid, and are not often seen because they haul out on the thickest ice where ships and researchers seldom venture. Like other seals, they are highly dependent on pack ice as a resting place for giving birth and for when they molt.

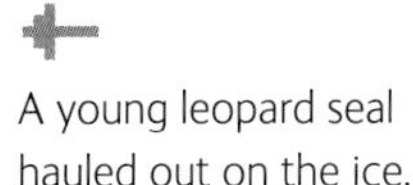

A young leopard seal hauled out on the ice.

THE ELEPHANT OF THE SOUTHERN OCEAN Much more obvious are the colossal southern elephant seals, *Mirounga leonine*, which haul out on the northern beaches of the Antarctic Peninsula and the subantarctic islands, in particular South Georgia and the Macquarie islands. Males can balloon to 8140 pounds (3700 kg), as much as an Asian elephant, while females weigh in at a more slender 660–1880 pounds (300–850 kg). They spend 90% of their lives at sea, their immense blubber reserves crucial to their survival in the icy waters. Despite their bulk, elephant seals are superb divers, reaching depths of 4950 feet (1500 m) in two-hour dives looking for fish and their favorite – squid. The South Georgia populations of southern elephant seals alone are thought to eat about 3.9 million tons (3.5 million metric tons) of squid each year. Elephant seals were thought to generally avoid pack ice of any description, especially when young. However, recent satellite tracking from the Antarctic Peninsula has shown that adult female elephant seals can spend long periods around pack ice, feeding on Antarctic silverfish.

Southern elephant seals can become entangled in fierce battles over territories and harems.

The biggest of them all

The largest organisms to roam the Southern Ocean, indeed on Earth, are the whales. These majestic animals are a favorite with many people, but some species are an increasingly rare sight. Commercial whaling (mostly between 1900 and 1960) slashed the blue and humpback whale populations to 1%. It is now illegal to whale commercially in the Southern Ocean, but this decimation cannot be reversed, so numbers of some species remain low. A prerequisite for all whales in ice-covered waters is that there are sufficient areas of open water for them to surface and breathe. Consequently, few whales are found deep within the pack ice during winter, and typically they only migrate to feed in southerly parts of the Southern Ocean in the ice-free summer months.

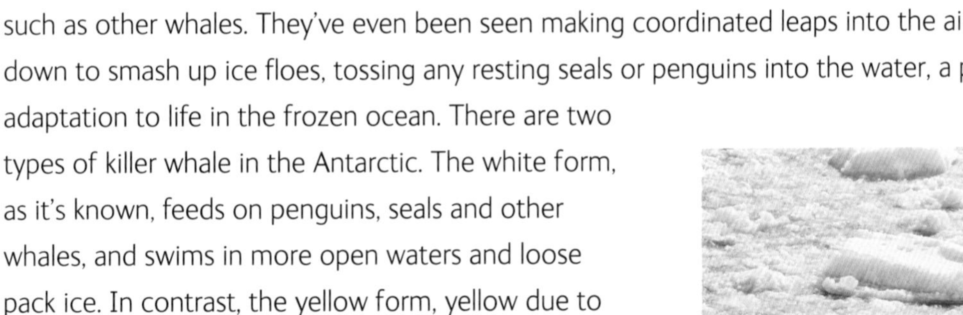

KILLERS OF THE PACK Besides the leopard seals, killer whales, *Orcinus orca*, are the top predatory animals. They typically hunt in pairs, or on occasions in pods of up to 50 individuals, around the outer margins of the pack ice. By feeding in pods, killer whales can tackle large prey such as other whales. They've even been seen making coordinated leaps into the air, coming down to smash up ice floes, tossing any resting seals or penguins into the water, a phenomenal adaptation to life in the frozen ocean. There are two types of killer whale in the Antarctic. The white form, as it's known, feeds on penguins, seals and other whales, and swims in more open waters and loose pack ice. In contrast, the yellow form, yellow due to a covering of small microscopic organisms called diatoms, feeds mainly on fish deeper in the pack ice.

Orcas travel within the pack ice but need cracks in the ice through which to come up for air.

Sperm whales, *Physeter macrocephalus*, are the largest of the toothed whales and cruise the outer margins of the Antarctic pack-ice zone. These whales feed almost exclusively on squid, and they can dive for periods of up to two hours and to depths of 9850 feet (3000 m). Sperm whales have a characteristic bulbous

BLOG:

I awoke to a glorious sunny day and was treated to my first glimpse of a whale in many days. It appeared in a small lead, less than 1000 feet from the ship. By 10:00 am I was unexpectedly in the helicopter ... being transported to an ice floe ... It was amazing, for the first time I could sit by a lead and listen to the new ice move and crackle.

lump on their heads, called a melon. Filled with a semiliquid waxy substance called spermaceti, the precise use for the melon is subject to much debate. One theory suggests it's used in echolocation, bouncing sound signals off objects so the whales can orient themselves, a bit like a bat. Others suggest they use the sound signals to stun squid or other prey. Certainly their prey are very fast and sparsely distributed in the great ocean depths. Anything to help catch up with them would be very useful.

THE BALEEN WHALES Humpback, *Megaptera novaeangliae*, southern right, *Eubalaena australis*, sei, *Balaenoptera borealis*, fin, *Balaenoptera physalus*, and blue whales, *Balaenoptera musculus*, all travel huge distances in summer to feed in the ice-free Southern Ocean. They are the biggest species there, but they feed on the smallest: tiny plankton, in particular krill. These baleen whale species have a curtain of fine plates called baleen, made from keratin, that hang down from the roof of the mouth. The plates act like giant filters to sieve out the plankton.

These whales are rarely seen near the pack ice, and if so only at the very outer margins. They feed predominantly on krill swarms and have various strategies for catching the vast quantities they need. Humpback whales release curtains of bubbles around the krill and then lunge up from beneath to gulp them up. Southern right whales skim the surface of the ocean for food with their jaws wide open, and blue, fin, humpback and minke whales take huge gulps of krill-rich water before closing their mouths and forcing the water out through the baleen plates. This leaves behind a thick plankton soup they then swallow. All these large krill feeders need to catch about 3% of their body weight in food each day, which for a 112-ton (100 metric ton) animal means a massive 3.4 tons (3 metric tons) of crustaceans.

The only baleen whales to regularly move deep within the cracks and breaks in the pack ice are the Antarctic minke whales, *Balaenoptera bonaerensis*, a spectacular sight as they gently arch between ice floes. Krill is again on the menu. Because minke whales have only very small dorsal fins near the tail, they can use their back to ram thin ice to break it up to produce their own breathing holes.

A humpback whale diving in shallow coastal waters.

BLOG:

THE CORERS ARE CLEANED AND PACKED AS IS MY OVERSIZED SUIT AND FOR NOW I MUST BE CONTENT WITH THE INCREDIBLE EXPERIENCE THAT I HAVE HAD ... Unlike those mulling over the "will I, won't I return" question I have no doubt that I will. The work has been hugely enjoyable and rewarding even if it has been at times physically demanding and the hours long.

• • •

Will Antarctica survive?

Climate change

MENTION YOU'RE OFF TO THE ANTARCTIC and very quickly people start talking about climate change, melting ice caps and the ensuing rise in sea level. The film *The Day After Tomorrow* did much to fuel a vivid, albeit fanciful, impression of the horrors that may follow. Phrases such as "polar meltdown" are bandied about as tales of disintegrating ice shelves are reported in the press. The best estimates are that within the next 100 years average global temperatures will have increased by about 4°F (2°C), and the effects of these increases will be most apparent in high latitudes. But even here it is predicted that some areas will get colder and some warmer.

Icicles adorn an ice shelf warming in the summer sunshine.

With all these predictions flying around, it's sometimes difficult to make sense of what is really happening. One of the reasons for the apparent confusion is that in the media ice sheets often get mixed up with pack ice or ice shelves or even icebergs. This is compounded by the way Antarctica is often seen as a whole rather than a complex series of different systems. Coupled with this is the fact that the scientists trying to make the pertinent measurements are dealing with rather limited data sets of reliable measurements. The ice core and ocean sediment records tell us about the enormous cyclical behavior of the Earth's climate over long geological time scales, and it's hard to interpret a few decades of data in the much larger scheme of things.

A WARM ANTARCTIC PENINSULA We do know that the Antarctic Peninsula has warmed by about 4.5°F (2.5°C) over the past 50 years. In the same period, many of the glaciers have begun to retreat by an average of 165 feet (50 m) per year over the past five or six years. There has also been a loss of about 3100 square miles (8000 sq km) of ice shelves since the 1950s, and scientists believe both are linked to temperature increases. The only question is whether or not the temperature increases are due to the activities of humans or are part of the result of climatic events that would have happened anyway.

Ice cores reveal priceless information about the snow and, therefore, atmosphere from which the ice is derived.

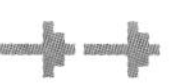

Samples of glacial ice are ready for transport back to the laboratory for sophisticated chemical analyses.

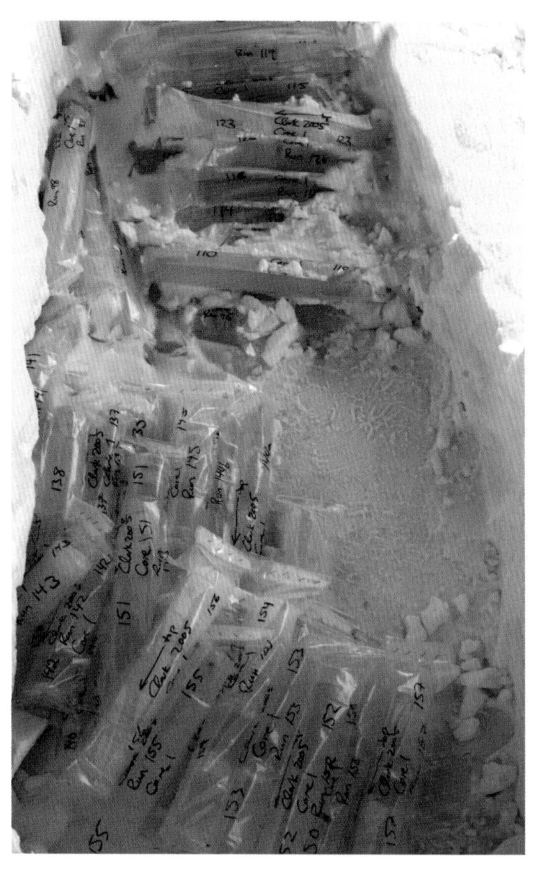

MELTING OF ICE SHEETS Even though we can't know the exact future, it's unlikely all the Antarctic ice sheets will melt, releasing their water to cause a catastrophic rise in sea level. At least not in the next 100 to 200 years. If the climate over the ice caps does warm up, what's more probable is greater snowfall, which in turn will lead to more ice being deposited. As temperatures increase there is more precipitation, and in cold places like the Antarctic the moisture falls as snow.

Some measurements show the West Antarctic Ice Sheet is thinning, while the East Antarctic Ice Sheet is thickening. There are some reports that say the two sheets are in a sort of balance, and losses from the west are counterbalanced by gains in the east. However, most recent studies indicate this isn't the case, and that the rate of thinning in the west is far greater than the rate of gain in the east. Debate is ongoing, and we'll only discover the truth through

Digging a hole in the snow is a useful freezer for storing precious samples.

intensive research and investment in new scientific methods to take accurate enough measurements. This is not easy on such large masses of ice.

PACK ICE AND WARMING There has actually been a slight overall increase in Antarctic pack-ice extent over the past 30 years, while overall annual pack-ice extent in the Weddell Sea, Indian Ocean and Pacific Ocean appears fairly stable. And while the Bellingshausen and Amundsen seas in the west have lost about 8% per decade, it seems that there has been a corresponding increase in ice cover of about 7% per decade in the neighboring Ross Sea.

Having said that, maximum pack-ice extent during the summer has actually decreased by about 3.7% per decade, although these decreases are considerably less than in the Arctic. However, such generalizations mask significant regional differences. Ice seasons in the eastern

Numbers of tourists are growing, although where they can go is strictly controlled.

Ross Sea, far western Weddell Sea, far eastern Weddell Sea and coastal regions of the east Antarctic have shortened. In contrast, pack ice seasons have lengthened in the western Ross Sea, Bellingshausen Sea and central Weddell Sea.

IT'S WORSE IN THE ARCTIC Whatever the effects of global climate change in the Antarctic, things are worse at the other end of the Earth, in the Arctic. It's significantly warmer for a start, and there is very convincing data that the ice sheets and glaciers of Greenland are melting at an increasingly rapid rate. In 1996 the loss was estimated to be around 25 cubic miles (100 cubic km) per year. Less than ten years later that loss had doubled.

Climate has had a much more obvious effect on the Arctic pack ice. Satellite data have shown that Arctic pack ice lasting through the summer has declined by 9% each year since 1978. If this rate continues, there might not be any pack ice in the summer by the end of the century. Other studies have shown the pack ice in the Arctic is considerably thinner now than it was in the 1950s, a trend particularly noticeable in the past decade.

Visitor pressure

The immediate effect of tourism on Antarctica is much easier to predict. In 1987 only about 1000 people visited. By 2003 there were nearly 20,000 tourists, and in 2006 that number was nearer 28,000. It's not crisis point yet, however. Antarctica is a huge place, and although the number of visitors has increased, the number of ships entering Antarctic waters and planes landing on the continent remain very few indeed.

Since there are few icebreaker tourist boats, they can only operate in the very edges of the pack ice. The tourist itineraries are also normally based on coastal bases, and visits are only made during the summer. Tour operators actively manage trips to Antarctica under very strict guidelines through the International Association of Antarctica Tour Operators (IAATO). For instance, there are very strict regulations about how close anyone can get to hauled out seals and penguin colonies, rules about waste disposal and even rules about disinfecting footwear before getting on and off the ship.

PROTECTED AREAS Most of the tourist activity is still concentrated in the Antarctic Peninsula, but operators are becoming more adventurous and customers more willing to pay for that unforgettable experience. It's this ambition to go farther and see more that must be watched carefully. Trips to inland sites are on the increase and over time will become more affordable. In advance of this, the Antarctic Treaty signatories have implemented a scheme for designating certain areas Antarctic Specially Managed Areas (ASMA). The first ASMA was outlined in 2004 and includes the Antarctic dry valleys. By managing all human activities, both the scientist's and the tourist's, the scientific, wilderness, ecological and aesthetic values of the valleys will be protected. It's hoped more ASMA agreements are signed, and soon. There can be no doubt tourism in the Antarctic is a positive thing provided it is done in a way that is ecologically sustainable and socially responsible.

FOOTPRINTS The effect tourists have on Antarctica, despite even the best intentions, are best described in David G. Campbell's award-winning book *Crystal Desert*. In it, he recounts meeting a group of tourists, and in particular one older lady who wanted to "see Antarctica, and the uncluttered face of the world" before she died. He watched her one day as she went to photograph a whale skeleton, trampling along the way a bed of moss around the bones. Later she climbed to see another site and missed her footing. She dislodged several rocks and lichens as a result. Her visit was only brief, but despite appreciating what she saw she certainly left her mark.

The Antarctic is a hostile place, but its beauty is simply staggering.

It's not just tourists that leave a mark. Scientific teams have not always been good at cleaning up after themselves. Huge efforts are now making up for this, to clean out the remains of past and less careful times. True, there are now stringent rules to control what is taken in and what is taken out. However, in the taking of one large core from the ice sheet, for example, there is inevitably a large environmental footprint left behind. The good news is that, due to the Antarctic Treaty, an assessment of the environmental impact of any activity in the Antarctic is scrutinized well before it can take place, which is an essential part of planning your work. Despite such precautions, some research camps operate for several years, belching out exhaust from engines and generators. Drilling rigs use tons of fluids to stop the corers freezing, and no matter how careful a field party, their footprints will always leave a record of their visit, especially as they open up ancient lakes and pristine study sites. This is the cost of progressing our understanding. It is always worth considering that our quest for knowledge comes at a tremendous cost to the system we are so keen to find out about.

It's a paradox that one of the most hostile places on the planet comprises fragile ecosystems so easily disturbed and wrecked. Life in the Antarctic, be it a lichen or an emperor penguin in the heart of winter, is really just hanging in there at times. It is all the more fascinating, and more beautiful, as a result.

More about Antarctica

Books

Antarctica: The Blue Continent, L. Woodworth. Frances Lincoln, 2005.

Antarctica: The Complete Story, D. McGonigal and L. Woodworth. Frances Lincoln, 2003.

Antarctica: A Guide to the Wildlife, T. Soper and D. Scott. Brandt Travel Guides, 2000.

Biology of Polar Habitats, The, G. E. Fogg. Oxford University Press, 1998.

Biology of the Southern Ocean, The, G. A. Knox. Cambridge University Press, 1994.

Complete Guide to Antarctic Wildlife, A, H. Shirihai and illustrated by B Jarrett. Princeton University Press, 2002.

Crystal Desert: Summers in Antarctica, The, D. G. Campbell. Mariner Books, 2002.

Explorations of Antarctica: The Last Unspoilt Continent, The, G. E. Fogg & D. Smith. Cassell Publishers, 1990.

Frozen Oceans, D. N. Thomas. Natural History Museum, 2004.

History of Antarctic Science, A, G. E. Fogg. Cambridge University Press, 1992.

Ice, The, S. J. Pyne. Weidenfeld & Nicolson, 2003.

Ice Blink: An Antarctic Essay, S. Faithfull. Book Works, 2006.

Life in Ancient Ice, J. D. Castello & S. O. Rogers. Princeton University Press, 2005.

Life in the Freezer, A. Fothergill. BBC Worldwide, 1993.

Lonely Planet: Antarctica (Country and Regional Guides), J. Ruben. Lonely Planet Publications, 2000.

Lonely Planet: Antarctica – A Travel Survival Kit, J. Ruben. Lonely Planet Publications, 1999.

March of the Penguins, L. Jacquet. National Geographic Books, 2006.

Terra Incognita: Travels in Antarctica, S. Wheeler. Vintage, 1997.

Under Antarctic Ice: The Photographs of Norbert Wu, J. Mastro. University of California Press, 2004.

Websites

N.B. Website addresses are subject to change. There are many websites that deal with Antarctica. This is not intended to be an exhaustive list, but rather something to start you off.

Alfred Wegener Institute for Polar and Marine Research **www.awi.de**
Antarctic Heritage Trust **www.heritage-antarctica.org**
Antarctica New Zealand **www.antarcticanz.govt.nz**
Australian Antarctic Division **www.aad.gov.au**
British Antarctic Survey **www.antarctica.ac.uk**
Cool Antarctica **www.coolantarctica.com**
International Polar Year 2007-2008 **www.ipy.org**
NASA **www.nasa.gov**
NASA Earth Observatory **www.earthobservatory.nasa.gov**
NASA, JPL Oceanography Group Oceans **www.jpl.nasa.gov/polar**
National Oceanic and Atmospheric Administration **www.noaa.gov**
National Institute of Polar Research, Japan **www.nipr.ac.jp**
National Science Foundation Polar Programme **www.nsf.gov/dir/index.jsp?org=OPP**
National Snow and Ice Data Center **www.nsidc.org**
Natural History Museum, London **www.nhm.ac.uk**
www.nhm.ac.uk/nature-online/earth/antarctica
www.nhm.ac.uk/nature-online/earth/oceans/frozen-oceans/frozen-oceans-weekly-journals.html
School of Ocean Sciences, Bangor University **www.sos.bangor.ac.uk**
Scientific Committee on Antarctic Research **www.scar.org**
Scott Polar Research Institute **www.spri.cam.ac.uk**
The Antarctic Circle **www.antarctic-circle.org**
The Antarctic Sun **www.antarcticsun.usap.gov/2005-2006/sctn02-12-2006.cfm**
United States Antarctic Programme **www.usap.gov**

Index

Italic pagination refers to illustrations.

Picture credits

Front cover, back cover main, inside back flap, end papers © David N. Thomas; back cover map © Lisa Wilson/NHMPL; **pp.1, 2, 9, 11** © David N. Thomas; **p.12l** Photo by Alexander Colhoun/NSF; **p.12r** Photo by Melanie Conner/NSF; **p.13** © David N. Thomas; **p.15** © Mark Brandon; **p.16** Photo by Spencer Klein/NSF; **pp.17, 20** © Lisa Wilson/NHMPL; **p.21** Photo by Zee Stevens/NSF; **p.22** © Lisa Wilson/NHMPL; **p.23** Photo by Kristan Hutchinson/NSF; **p.25** © David N. Thomas; **p.26** Photo by Tracy Szela/NSF; **p.27** Photo by Patrick Rowe/NSF; **p.28** © Cornelia Thomas; **p.29** © David N. Thomas; **p.31l** Photo by Kristan Hutchison/NSF; **p.31r** Photo by Tracy Szela/NSF; **p.32** © David N. Thomas; **p.34** Photo by Kristan Hutchison/NSF; **p.35t** Photo by Josh Landis/NSF; **p.35b** © David N. Thomas; **p.38** © Nicola Dunn; **p.39** Photo by Forest Banks/NSF; **pp.40–44** © David N. Thomas; **p.45l** Photo by Emily Stone/NSF; **p.45m** Photo by Kristan Hutchison/NSF; **pp.45r, 46** © David N. Thomas; **p.47l** Photo by Nick Powell/NSF; **p.47r** © Louiza Norman; **p.48** Photo by Brien Barnett/NSF; pp.49, 50l © David N. Thomas; p.50r © Lisa Wilson/NHMPL (reproduced from www.gaisma.com/en); p.51 © David N. Thomas; **p.55** © Mark Brandon; **p.56** Photo by Emily Stone/NSF; **p.57** Photo by Glen Snyder/NSF; **p.58** Photo by Emily Stone/NSF; **p.59l** © David N. Thomas; **p.59r** © Jacqueline Stefels; **p.60** Photo by Kurtis Burmeister/NSF; **p.63** Photo by Henry Kaiser/NSF; **p.64** P. Marschall/Alfred Wegener Institute for Polar and Marine Research; **p.65t, m** Photo by Steve Clabuesch/NSF; **p.65b** Photo by Henry Kaiser/NSF; **p.67** © Paul Drummond/Arcticphoto.com; **pp.69–73** © David N. Thomas; **p.74** Photo by Katy Jensen/NSF; **p.75** © David N. Thomas; **p.76l** Photo by Steven Profaizer/NSF; **pp.76r, 77** © David N. Thomas; **p.78** Photo by Glenn E. Grant/NSF; **p.79** Photo by Jaime Ramos/NSF; **p.81** © Frank Todd/Arcticphoto.com; **p.85** © David N. Thomas; **p.86l** Photo by Emily Stone/NSF; **p.86r** Photo by Emily Stone/NSF; **p.87** © David N. Thomas; **p.88** © Mark Brandon; **pp.90, 95, 96** © David N. Thomas.

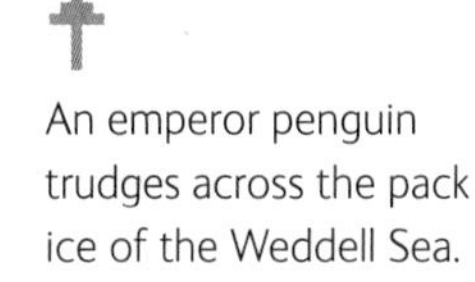

An emperor penguin trudges across the pack ice of the Weddell Sea.

Every effort has been made to contact and accurately credit all copyright holders. If we have been unsuccessful, we apologize and welcome correction for future editions and reprints.

(NSF, National Science Foundation; NHMPL, Natural History Museum Picture Library)

Author's acknowledgments

Once again I have to thank Cornelia Thomas for her support and patient understanding for me taking on this project while packing up to go to the Antarctic yet again. There are a host of experiences and opportunities that have led to me being able to attempt such a project, and I am immensely grateful to all those colleagues and friends over the past 15 years that have made these possible.

Publisher's acknowledgments

We would like to thank Antarctic Heritage Trust for giving permission to reproduce selected blogs from those written by Nicola Dunn, Robert Clendon, Sarah Clayton and Ainslie Greiner during their time in Antarctica in 2006, conserving artifacts from Ernest Shackleton's hut left during his team's expedition to Antarctica in 1908. The Antarctic Heritage Trust is an independent charitable trust based in Christchurch, New Zealand. It was formed in 1987 to care for the heritage of the heroic era located in the Ross Sea region of Antarctica on behalf of the international community.
www.piclib.nhm.ac.uk/antarctica/

We would also like to thank David N. Thomas, Stathys Papadimitriou and Louiza Norman for giving permission to reproduce selected extracts from the online diary of their 2006 expedition on board the German *RV Polarstern* to the sea ice of the Weddell Sea in the Antarctic.
www.sos.bangor.ac.uk/antarctic.htm